Ken Ford has written a truly remarkable book. It is not only the story of his life but it is a primer for the nuclear age. He is one of the few witnesses left of how the hydrogen bomb was created. There are portraits of people like Edward Teller and John Wheeler and the physics is clear. It is a must read for anyone who wants to learn about the history of nuclear weapons.

Jeremy Bernstein, author of *Nuclear Iran* and *A Chorus of Bells and Other Scientific Inquiries.*

Ken Ford's book provides a first-hand look at the early days of U.S. thermonuclear weapons design and the work under John Wheeler of the Matterhorn B (for bomb) project at Princeton that focussed on predicting the yield of the first U.S. test of a hydrogen bomb—Mike—on November 1, 1952. Knowing all the participants, I found the account accurate as well as entertaining.

Richard Garwin, principal designer of the Mike thermonuclear device, co-author of *Megawatts and Megatons: The Future of Nuclear Power.*

Building the H Bomb offers a rare and fascinating insider's look at the making of humankind's most powerful nuclear weapon. Ford combines his trademark talent for explaining physics with a warm, engaging personal story. Amidst the darkness of Cold War paranoia and the nuclear arms race, Ford lets the scientists' personalities, their quest for knowledge, and his own youthful innocence shine through. Part physics, part history, part memoir, this book reminds us that science is ultimately a very human endeavor.

Amanda Gefter, author of *Trespassing on Einstein's Lawn*

A charming and engrossing book about the building of the hydrogen bomb, the individuals who built it, the development of computers and much more from the point of view of a young man who was in the middle of all of it and knew everybody.

Gino Segrè, author of *Faust in Copenhagen* and *Ordinary Geniuses,* at work on a biography of Enrico Fermi.

The development of the hydrogen weapon that was carried out simultaneously in the USA and the USSR is a wonderful example of dedication, professionalism and brilliant work of scientists in our two countries.

Publications by first-hand participants of these projects are becoming especially precious. This book is a rare example of first-hand recollections of a participant providing popular, but precise description of the H-bomb invention process.

Ford's recollections of meetings with many outstanding scientists of the 20th century who have turned into historical figures by now, including those who were very young at that time—for example, Richard Garwin—are especially valuable.

The combination of clear description of physical processes and vivid personal emotions with precise depiction of political climate in the USA at that time is of special value.

It was a pleasure to read this book and I wish many would read it.

Zhores I. Alferov, rector of St. Petersburg's Academic University, winner of the 2000 Nobel Prize in Physics, and the subject of Paul Josephson's *Lenin's Laureate: Zhores Alferov's Life in Communist Science.*

Building the
H Bomb
A PERSONAL HISTORY

Building the H Bomb

A PERSONAL HISTORY

Kenneth W. Ford

NEW JERSEY · LONDON · SINGAPORE · BEIJING · SHANGHAI · HONG KONG · TAIPEI · CHENNAI

Published by

World Scientific Publishing Co. Pte. Ltd.
5 Toh Tuck Link, Singapore 596224
USA *office*: 27 Warren Street, Suite 401-402, Hackensack, NJ 07601
UK *office*: 57 Shelton Street, Covent Garden, London WC2H 9HE

Library of Congress Cataloging-in-Publication Data
Ford, Kenneth William, 1926–
   Building the H bomb : a personal history / Kenneth W Ford.
     pages cm
   Includes bibliographical references and index.
   ISBN 978-9814632072 (hardcover : alk. paper) -- ISBN 978-9814618793
   (pbk. : alk. paper)
   1. Hydrogen bomb--Design and construction--History. 2. Hydrogen bomb--
   United States--History. 3. Ford, Kenneth William, 1926–   I. Title.
   UG1282.A8F67 2015
   623.4'5119--dc23

                                            2014029386

British Library Cataloguing-in-Publication Data
A catalogue record for this book is available from the British Library.

First published 2015
Reprinted 2015

Book design and layout by Adam B. Ford
Fonts used are Lora, by Olga Karpushina and Alexei Vanyashin, available from cyreal.org, Planscribe NF, by Nick Curtis, available at nicksfonts.com, Proxima Nova, by Mark Simonson, available from marksimonson.com, and Special Elite, by Brian J. Bonislawsky, available from astigmatic.com.

Cover images are from the cover of a classified scientific paper by US nuclear physicists Edward Teller and Stanislaw M. Ulam, March 9, 1951, *On Heterocatalytic Detonations I: Hydrodynamic Lenses and Radiation Mirrors*, at US nuclear weapon design facility Los Alamos Scientific Laboratory, Los Alamos, New Mexico.

Required statement regarding the cover images: Unless otherwise indicated, this information has been authored by an employee or employees of the University of California, operator of the Los Alamos National Laboratory under Contract No. W-7405-ENG-36 with the U.S. Department of Energy. The U.S. Government has rights to use, reproduce, and distribute this information. The public may copy and use this information without charge, provided that this Notice and any statement of authorship are reproduced on all copies. Neither the Government nor the University makes any warranty, express or implied, or assumes any liability or responsibility for the use of this information.

Printed in Singapore

To the memory of my many nuclear friends
who are no longer with us

Table of Contents

Preface · ix
A Note on Secrecy · · · · · · · · · · · · · · · · · · · x
Prologue · xi
1. The Big Idea · 1
2. The Protagonists · · · · · · · · · · · · · · · · · 13
3. The Choice · 24
4. The Scientists, the Officials, and the President · · 37
5. Nuclear Energy · · · · · · · · · · · · · · · · · · 44
6. Some Physics · · · · · · · · · · · · · · · · · · · 59
7. Going West ·71
8. A New World · · · · · · · · · · · · · · · · · · · 81
9. The Classical Super · · · · · · · · · · · · · · · · 92
10. Calculating and Testing · · · · · · · · · · · · · ·106
11. Constructing Matterhorn · · · · · · · · · · · · 119
12. Academia Cowers · · · · · · · · · · · · · · · · ·132
13. New Mexico, New York, and New Jersey · · · · ·140
14. The Garwin Design · · · · · · · · · · · · · · · ·152
15. Climbing Matterhorn · · · · · · · · · · · · · · ·163
16. More Than a Boy · · · · · · · · · · · · · · · · · 177
Epilogue ·185
Acknowledgments · · · · · · · · · · · · · · · · · ·189
End Notes · 191
Bibliography · 205
Index · 209

Preface

This book is three things. It is a history of the development of the world's first hydrogen bomb. It is also a memoir of the author when he was twenty-four to twenty-six years old. And it is a mini-textbook on nuclear physics. Think of it as a three-stranded braid, not a three-ingredient blend. As you proceed, you will encounter bumps here and there as one strand yields to another. I have tried to lighten and brighten the trip with lots of illustrations. I hope that in the end you will conclude that it all hangs together and makes for an agreeable as well as informative ride.

Ken Ford

A Note on Secrecy

According to the United States Department of Energy, this book contains some secrets. I disagree.

During the period that is the focus of this book, 1950-1952, I held what was (and still is) called Q clearance. That gave me access to secret and even some top secret weapons information, and it gave me the opportunity to create such information. In this book, in addition to personal recollections of people and events from that period, I discuss weapons information that may have once been secret but is now in the public domain. I have bent every effort to avoid revealing any information that is still secret.

Any technical details that I provide, such as dimensions or weight or energy yield, are taken from now publicly available sources, not from my very hazy memory (with *one* exception—see page 177). It's been more than sixty years!

In my considered opinion, this book contains nothing whose dissemination could possibly harm the United States or help some other country seeking to design and build an H bomb.

Ken Ford

Prologue

In the summer of 1942, three years before nuclear weapons devastated Hiroshima and Nagasaki, a small group of notable physicists gathered in Berkeley, California. Their mission: to consider how nuclear physics could be applied to war. They knew that there were two possible ways in which atomic nuclei might release immense energy, energy vastly greater than could be provided by the kinds of explosive weaponry then in use. Those two ways were *fission* and *fusion*.

The energy of "conventional" weapons is *chemical* energy. It is achieved when individual atoms rearrange their links to other atoms in chemical reactions. *Nuclear* energy, by contrast, is achieved, as its name suggests, when the tiny cores at the centers of all atoms—the atomic nuclei—are rearranged in nuclear reactions.

Nuclear fission was discovered in December 1938. A nucleus of the heavy atom uranium (number 92 in the periodic table), when struck by a small electrically neutral particle called a neutron, can break apart into two fragments, each of which is the nucleus of an atom much lighter than uranium. In the process energy is released, far more energy per atom than in chemical change. (The name fission was borrowed from the terminology for cell division in biology.)

But vast energy per atom means little if there are not trillions of trillions of atoms involved. That's where the chain reaction comes in. The year 1939 had hardly begun before physicists—first in the United States, then around the world—learned that in each fission process, on average, two or more additional neutrons are emitted, opening the possibility that one fission event could trigger two more, which could then trigger four, then eight, and so on until, in a small fraction of a second indeed trillions of trillions of nuclei could undergo fission, and devastation on an unheard-of scale could be the result.

Nuclear fusion was known as a laboratory phenomenon many years before nuclear fission was discovered, but it was not until 1939 that physicists fully embraced the idea that the energy of the Sun—and other stars—comes from the combination (or "fusing") of the nuclei of hydrogen (number 1 in the periodic table) to form nuclei of the heavier element helium (number 2). Just as with fission, this fusion process releases far more energy per atom than does chemical change. And, just as with fission, this fact means little unless an astronomical number of atoms take part in a very short time.

This knowledge of fission and fusion is what the physicists in Berkeley had before them that summer. It took little time for them to conclude that a fission bomb (or A bomb, as it came to be called) was very probably feasible. By that fall, the Manhattan Project would be officially launched, and by the next spring work on fission bombs would shift into high gear—work that culminated, in the summer of 1945, in the successful "Trinity" test in the New Mexico desert and in the still-controversial decision to drop fission bombs on Japanese cities. (By that time, plutonium had been added to the roster of fissionable elements, and two of the three nuclear explosions in 1945 used that element.)

The Berkeley physicists were intrigued by the possibility of a fusion bomb, but much less sure of its feasibility. As a result, it was not accorded a high priority during the war years. As of 1945, work on the H bomb (as the fusion bomb came to be called*) had not led to a brighter prospect for its success. If anything, the work in the intervening years made the prospect

---

*The H bomb, or hydrogen bomb, is also called a thermonuclear weapon, because its operation requires high temperature—*extremely* high temperature.

dimmer. Nevertheless, work continued, for the H bomb, if it could be made to work, would be far more powerful than an A bomb, and might be less costly to make. For those concerned with "more bang for the buck," it was an attractive option. For some others, it was almost too horrendous to imagine. Yet nearly all the physicists and policy makers agreed that work to establish its feasibility (or not) should continue.

As of 1950, when I joined the effort, there had been no breakthrough, and the likelihood of success in building an H bomb was at best clouded. Then, in the spring of 1951, came an idea that *was* a breakthrough. That is where my story begins.

Chapter 1
The Big Idea

O n March 9, 1951, Edward Teller and Stan Ulam issued a report, LAMS 1225*, at the Los Alamos Scientific Lab[†] where they both worked at the time. It bore the ponderous, hardly illuminating title "On Heterocatalytic Detonation I. Hydrodynamic Lenses and Radiation Mirrors," and it changed everything. Since it dealt with thermonuclear weapons (H bombs), it was, of course, classified secret. For some reason, it remains secret to this day. The highly redacted version of it that can be found on the Web[1] is mostly white space. Nevertheless, most of what was in it is well known.

Their big idea, which we refer to now as radiation implosion, was that the electromagnetic radiation (largely X rays) emitted by a fission bomb, if appropriately channeled, could compress and heat a container of thermonuclear fuel sufficiently that that fuel would be ignited and the nuclear flame would propagate, not fizzle. The expected result: megatons of energy, not kilotons.[‡] History validated the Teller-Ulam idea.

---

*As an LAMS report (MS for manuscript), it was lesser ranked (or more informal) than an LA report.

[†]Now the Los Alamos National Laboratory.

[‡]A "ton" of energy is the nominal energy released when one ton of high explosive blows up (for the record, it is 4.2 billion joules). The energy released in nuclear explosions is measured in thousands or millions of tons (kilotons or megatons). The Hiroshima bomb, the first nuclear weapon used in war, had an estimated "yield" of 13 to 15 kilotons. The second weapon, dropped on Nagasaki, yielded about 21 to 23 kilotons. In subsequent tests, the yields have been measured with greater precision. To put all of this in human terms, a ton of explosive energy is about the same as a million food calories, enough to keep a human going for about 500 days. A kiloton would "feed" a thousand people for 500 days. A megaton, spread over that same period of time, would nourish a million people. But that same megaton, released in a fraction of a second in the right place could slaughter a million people.

Stan Ulam, 1951. *Courtesy of AIP Emilio Segrè Visual Archives, Physics Today Collection.*

(In the end, it was even more effective than they first imagined.) On exactly who contributed what to that big idea, history is a little fuzzier. More on that below. (Here and in what follows, I use "Teller-Ulam" not to anoint Teller as the senior author but only to keep the authors in alphabetical order, as they are on the report's cover.)

Stanislaw Ulam (always known as Stan) and Edward Teller (always Edward, never Ed) had some things in common. They were both émigrés from Eastern Europe—Stan from Poland, Edward from Hungary. They were both brilliant. They both had great curiosity about the physical world. And they were both a bit lazy. But oil and water also have some things in common. Stan and Edward differed more than they were alike. Stan, a mathematician with a gift for the practical as well as the abstract, was—to use current slang—laid-back. He had a droll sense of humor and a world-weary demeanor. He longed for the Polish coffee houses of his youth and the conversa-

2

Edward Teller, 1951. *Courtesy of AIP Emilio Segrè Visual Archives, Gift of Carlo Wick.*

tions and exchanges of ideas that took place in them. Edward was driven—driven by fervent anticommunism, by a desire to excel and be recognized—driven, it often seemed, by internal demons. Edward was too intense to show much sense of humor. Stan had an abundance of humor. Stan and Edward did not care very much for each other (which may help to explain why a "Heterocatalytic Detonation II" report never appeared).

I was a twenty-four-year old junior physicist on the H-bomb design team at Los Alamos when the Teller-Ulam report was issued. I saw Stan and Edward every day. I liked them both, and continued to like them, and to interact with them now and then, for the rest of their lives. Stan and I later wrote a paper together, on using planets to help accelerate spacecraft (the so-called "slingshot effect"). Edward and I later worked together as consultants to aerospace companies in California.

Not everyone at the lab had equal affection for these two men. Carson Mark, the Canadian mathematician turned research administrator who headed the Theoretical Division during the H-bomb period, could scarcely abide Edward. He liked Stan, even if Stan didn't care much for bureaucratic niceties and even if Stan sometimes wanted to chat when Carson wanted to work. John Wheeler, my mentor, although a straight-arrow quintessential American (he was born in Florida and raised in California, Ohio, and Maryland), was Edward's soul mate. They were completely in tune in their anti-Communism and their fear of Soviet aggression. Balancing their pessimism about world affairs, they shared an optimism that nature would, in the end, abandon all resistance and yield her secrets if they just pressed hard enough. They had done some joint research together back in the 1930s (on the rotational properties of atomic nuclei) and their wives, Mici (MITT-cee) Teller and Janette Wheeler, were friends. It was Edward's persuasion, in large part, that led Wheeler to interrupt a sabbatical in France and take a leave of absence from his academic duties at Princeton to spend the 1950-51 year at Los Alamos. Wheeler didn't exactly dislike Stan, he just didn't resonate with him. (There were, in fact, very few people whom Wheeler didn't like, and he tried hard to mask whatever negative feelings he had toward those few.) For Wheeler's taste, Stan was just a bit too laid-back, a bit too nonchalant.

Looking back, the odd thing to me now is that the Teller-Ulam idea, at the time it was advanced, didn't shake the Earth under our feet. There were vibrations, but no earthquake. There was a new sense of cheer, but no parties or toasts or flag waving. We didn't take the trouble to analyze, as so many have since, who exactly had what part of the idea and who deserves the greater credit. Years later, Edward said to me (I paraphrase), "Stan had a dozen ideas a day. They were almost all crazy. He himself had no idea which ones were valuable. It took me to pick out of the jumble the one good idea and exploit it." Also years later, Stan said to me (again, I paraphrase), "Edward just couldn't bring himself to admit, after his years of effort, that the idea on how to make the H bomb work was mine. He just had to take it and call it his own."

The Teller-Ulam idea landed in the midst of numerous other ideas, of varying complexity and varying chance of succeeding. These included "boosting" (having a small container of thermonuclear fuel at the center of a fission bomb to "boost" the fission bomb's yield); "Swiss cheese" (having numerous pockets of thermonuclear fuel scattered throughout fission fuel); the "alarm clock" (a name Edward Teller and Robert Richtmyer had coined in 1946[2] [3] for alternating layers of fission and fusion fuel,* and which Andrei Sakharov in the Soviet Union, as we later learned, had separately envisioned and separately christened a "layer cake" in 1948[5]); and the "Yule log" (John Wheeler's macabre name for a cylinder of thermonuclear fuel with no limit on its length or on its explosive power). Behind these lay the basic idea that had been around for nearly a decade and on which we were working assiduously at the time. That idea, known as the "Super" (and later as the "classical Super") was simple in concept but maddeningly difficult to model mathematically, so that there was no sure sense of its potential. At the time of the Teller-Ulam idea, however, there were more reasons for pessimism than optimism about the prospects of the classical Super. Calculations†

---

*Fitzpatrick[4] gives the specific date August 31, 1946, for the conception of the alarm clock idea, a date that Teller reportedly remembered because his daughter Wendy was born on that day. This is charming, and, with a bit of a stretch, consistent with Teller's statement in his Memoirs that he and Robert Richtmyer devised the idea "during the summer of 1946." Teller gives Richtmyer credit for the name.

---

†Calculations at the time were carried out mainly by people wearing skirts and blouses operating Marchant, Monroe, and Friden calculators. They were called "computers." Understandably, we often called them computresses. The nearest thing to a modern computer was a modified IBM accounting machine known as a card-programmed calculator (CPC). Computers as machines with internally stored programs came later—but not much later. I will return in Chapter 9 to a discussion of some of the "Super" calculations that were carried out when computers were still people and in Chapter 15 to calculations carried out on the true ancestors of modern computers.

kept suggesting that igniting the fuel, even with a powerful fission bomb, and even with a good deal of highly "combustible" tritium mixed in, would not be easy, and that even if it were ignited, it would probably fizzle rather than propagate. A homeowner trying to get a fire started in a fireplace with wet logs and inadequate kindling can relate to the difficulty.

So the Teller-Ulam idea landed in our midst not as "just" another idea—it was special—but also not as a lone idea where there were none already. It was like a new sapling introduced into a nursery, not like a palm tree miraculously delivered into the desert. We thought, "Now there is an idea with merit," and we started exploring its consequences at once—without immediately abandoning other ideas. As it turned out, the more we calculated, the more promising the new idea looked. Within three months, it had become *the* idea and was endorsed by the General Advisory Committee of the Atomic Energy Commission as the route to follow.

Up until February 1951, when Ulam approached Teller with an idea about imploding thermonuclear fuel and Teller realized (or, as he later claimed, recalled[6]) that radiation was the best thing to do the imploding, everyone working on H-bomb design in the United States assumed that the Super would have to be a "runaway" Super, a device in which the temperature of the material would have to "run away from" the temperature of the radiation. Otherwise, it seemed, the radiation would soak up too much of the energy and there wouldn't be enough left to ignite the thermonuclear fuel and keep it burning. What could change this bleak prospect, Ulam and Teller realized, would be great compression of the material. It was this February meeting and its insight that led to the Teller-Ulam report of March 1951 and to the new direction in H-bomb design.

There were two insights that flowed from the Teller-Ulam discussion. The first was that thermal equilibrium—that is, having the matter and the radiation at the same temperature—could be tolerated if there was enough compression. Occupying less volume, the radiation would soak up less of the total energy. More energy would be left to heat the matter and stimulate its ignition and burning. Up until then, those of us working on the Super accepted the idea that thermal equilibrium would be intolerable because of the excessive "loss" of energy to radiation. And we accepted an argument Teller had made[7] that compression would not help. Teller had pointed out that although compressing the thermonuclear fuel increases its reaction rate, it also increases, and by the same factor, the rate at which the matter radiates away energy. So there was no net gain, he had argued, from compression. But that argument posits a runaway Super, which was our mindset at the time. Once equilibrium is established, matter is not "losing" energy to radiation, it is just exchanging energy with radiation, gaining as much as it is losing. If you jump into the North Atlantic, you lose energy because your temperature is higher than that of the water, and you will soon be drained of your energy. If instead you jump into a hot tub, energy flows equally back and forth between you and the water as you remain in equilibrium, and you can bask there all afternoon.

I have not found in the written record any sure evidence that Stan Ulam had in mind this insight about equilibrium when he came up with the idea that the thermonuclear fuel should be compressed. Nor do I remember him explicitly mentioning it at the time. Yet I have to assume that he *did* have it in mind. Otherwise there would have been no good reason to argue for compression. He had already done quite a lot of work on the runaway Super and knew its disconcerting inability to hang onto enough energy to keep a reaction going. He most likely knew, also, the Teller argument that for a runaway Super, compression would not help.

Teller, in the now-famous conversation with Ulam, apparently did realize very quickly, despite his earlier arguments to the contrary, that compression could be a key to success. In his memoirs, written many years later[8] (from which I quote in the next chapter), he says that Ulam's idea was "far from original" and that, for the first time he [Teller] didn't object to it.[9] He doesn't tell us why he didn't object, an odd omission given his previous rejection of the idea. In the same paragraph, in a further put-down, Teller says that Ulam did not actually understand why compression was a good idea.

Our understanding of this meeting is murky indeed despite the clarity of the conclusion that flowed from it. Did Ulam come in with a full understanding of why compression might be the key to success in designing an H bomb? We don't know. Had Teller ever seriously entertained the idea of compression before? We don't know. (In later writings, Teller claims to have had the idea before Christmas 1950[10] and also about February 1, 1951.[11] These claims are dubious, especially in light of his own account of the meeting with Ulam,[12] and in light of my own recollection that no breakthrough idea occurred before late February 1951.) What we do know is that out of the meeting came the successful idea of the "equilibrium Super," in which compression is so great that the huge amount of energy soaked up by radiation in equilibrium with matter is tolerable.

The second insight that came from the Ulam-Teller meeting was that radiation, at sufficiently high temperature, is a "substance" with remarkable properties.* It can flow like a liquid and then push like a giant steel piston. No, not like steel. Stronger than steel. This is not the radiation emitted later in the explosive process by a thermonuclear flame. This is the radiation created by, and flowing from, a fission bomb—radiation that can be channeled until it surrounds a container of thermonuclear fuel and then implodes it. There is at least

---

*I discuss the physics of radiation in Chapter 6.

agreement that this idea of "radiation implosion" was Teller's. Ulam came with the idea that mechanical forces from a fission bomb could do the compressing. Teller saw that radiation would be an even better agent of compression. (According to the independent analyst Carey Sublette, who has written extensively on nuclear weapons without benefit of security clearance,[13] the radiation is indeed the "agent" of compression, with most of the actual "push" being supplied by ablation (that is, boiling off) of the outside of the container of thermonuclear fuel. The radiation bath also creates a plasma of electrically charged nuclei and electrons in low-density material just outside this container, further augmenting the push.[14])

How much compression? It depends on what the thermonuclear fuel is and, in any case, is not publicly known. But it is huge. Ten-fold? Twenty-fold? A hundred-fold? A thousand-fold? It is enough to stimulate a high rate of thermonuclear reaction in the fuel while keeping the volume occupied by the burning fuel sufficiently small that the radiation confined within that volume can't soak up too much of the total energy.

We tend to think of solid matter as incompressible—hard stuff of fixed volume—but it does in fact shrink and expand. The girders of a steel bridge that together span a thousand feet in the summer may shrink to 999 feet, 4 inches on the coldest winter day. Imagine instead (if you can!) a *ten*-fold compression in each dimension. The bridge would be a hundred feet long and a girder that was a foot across would be reduced to a width of little more than an inch. A cube of the original steel two inches on a side would weigh a little more than two pounds—easy to hold in your hand. A cube of the same volume of the compressed steel would weigh more than a ton.

The physics of nuclear weapons concerns more than nuclear reactions. It also concerns the properties of matter at temperatures and pressures and densities light years removed from anything that can be tested in the laboratory. Edward

Teller and Stan Ulam were among those theorists whose in-
genuity allowed them to visualize and to calculate what would
go on at these extreme conditions. What makes this possible?
The physicists' knowledge that the laws of electromagnetism
and of mechanics, both classical and quantum, extend to
domains far beyond direct observation; and their understanding
that ultimately, no matter what the conditions, one is dealing with the same electrons and nuclei and photons as in the
"ordinary" world around us.

Scientists and engineers in the Soviet Union were not
far behind those in the United States in developing and testing nuclear weapons (actually, for deliverable thermonuclear
weapons, they were briefly ahead[15]). From August 1945, when
the Soviets launched a serious program to develop an atomic
bomb,[16] until the explosion of "Joe 1" (Joe for Joseph Stalin) in
August 1949, only four years elapsed. Without the espionage
of Klaus Fuchs (and some others) it would have taken longer.
But perhaps not a great deal longer. The USSR had a comple-
ment of nuclear scientists as competent as those in the United
States. And not just competent. Some, like Andrei Sakharov,
Vitaly Ginzburg, and Yakov Zeldovich, were as brilliant as the
best in the West.

What about radiation implosion? When was the Teller–Ulam idea duplicated in the USSR? Interestingly, not until 1954 [17] This was Sakharov's "third idea."[18] In his *Memoirs*,
Sakharov avoids revealing secret information about the Soviet
thermonuclear program by speaking enigmatically only of the
"first idea," the "second idea," and the "third idea." By now it is
well known what these ideas were.[19] The first idea, advanced
by Sakharov in 1948, was the *Sloika*, or "layer cake," a design
much like the "alarm clock" that Teller and Richtmyer had proposed in 1946, and, so far as is known, conceived without the

help of espionage.[20] As initially imagined, the layer cake was to have alternating layers of uranium-235 and deuterium.[21] Sakharov and his team pushed hard on *Sloika* and indeed succeeded in incorporating it into what was the world's first deliverable thermonuclear weapon,* detonated in August 1953.

The "second idea," which Sakharov attributes to his colleague Vitaly Ginzburg,[23] was advanced in the Soviet Union in 1949. It proposed replacing deuterium as the thermonuclear fuel by a particular form of the compound lithium hydride in which the lithium is—either wholly or in substantial part— the isotope lithium-6 and the hydrogen is "heavy hydrogen," or deuterium.† I will have more to say about this compound— lithium-6 deuteride, as it is often called—in Chapter 9. In the United States, it was proposed by Edward Teller as a thermonuclear fuel in 1947.[24] Airdrops of weapons containing this solid thermonuclear fuel occurred first in the Soviet Union in November 1955[25] and in the United States (over Bikini atoll) six months later, in May 1956.[26] The two countries were never far apart in their developments of thermonuclear weapons.

I've been told that some historians of science have asked the question: Why did it take so long for scientists in the United States to come up with the idea of radiation implosion? Nearly nine years elapsed between the first discussions of a possible H bomb in 1942 (see Chapter 9) and the Teller-Ulam idea that made it practical in 1951. Just as pointedly, one could ask the same question of the Soviet scientists, where Sakharov's "third

---

*It was a "thermonuclear weapon" in the sense that more than a token part of its energy—about 15 to 20 percent—was contributed by thermonuclear reactions.[22]

---

†Ginzburg (as I learned from Zhores Alferov) adopted the code name Lidochka for this compound. Lidochka is the diminutive of the Russian female name Lida, which in turn bears a similarity to the chemical formula LiD.

idea" of radiation implosion came in 1954, also after nearly a decade of work on thermonuclear weapons, including periods both before and after Klaus Fuchs passed along some significant information in 1948 (which I shall discuss in Chapter 8). Both questions have, I think, the same answer: inflexibility. The scientists in both countries were like horses wearing blinders. Each group was pursuing a particular path and could see ahead but not to the side. In the United States the view ahead was of the classical Super. In the Soviet Union it was of the layer cake. Only when the view ahead got cloudy—when, in the United States, it looked more and more like the classical Super wouldn't work, and, in the Soviet Union, it looked more and more like the layer cake could never reach into the megaton range—only then did the scientists cast off their blinders and look in new directions.

And you the reader, could ask me, the once-young scientist, why I didn't come up with the idea of radiation implosion myself. It's a fair question. To be sure, I was a junior member of the team, but I understood all the relevant physics and I had a supple mind. What I lacked was a sufficiently questioning mind. Like my colleagues at the time, I accepted the idea that the only hope of igniting and sustaining a thermonuclear flame was to have the temperature of matter "run away from" the temperature of radiation. That, as Teller and Ulam realized (after the outlook for the classical Super became bleak) and as Sakharov concluded (after the prospects of the layer cake dimmed), was not true.

Chapter 2
The Protagonists

Both Teller and Ulam later wrote and spoke (including to me) about their February 1951 meeting and about the genesis of the idea of radiation implosion. Teller consistently claimed sole credit for the breakthrough idea—although his statements about the time when this happened vary. He gives Ulam credit only for earlier calculations showing that the classical Super was unlikely to work. Ulam's contribution, he said, was forcing new thinking about how to make an H bomb.

Ulam has consistently claimed that the idea of a many-fold compression of the thermonuclear fuel was his—at the same time acknowledging that the idea of using radiation for this purpose was Teller's.

Teller readily admitted that he did not care much for Ulam ("I had developed an allergy to him," he wrote[1]), and he believed that the ill will was reciprocated. I would describe Ulam's attitude toward Teller more as bemused mild derision than as hostility. They were not soul mates.

I devote this chapter to things that Teller and Ulam later said about the critical 1951 ideas that led to the successful H bomb.

When Teller was anointed Father of the H Bomb following the successful thermonuclear test in late 1952, he did not back away from the name, for indeed he thought he deserved it. Yet at the same time he was embarrassed, for many of his colleagues did not appreciate the searchlight of fame being focused on a single person.* So he graciously wrote a lengthy article to share the credit, "The Work of Many People."[4] In that article he cites Ulam's "disquieting," then "discouraging" calculations on the classical Super. Regarding new ideas and eventual success with the equilibrium Super, he credits Ulam with "an imaginative suggestion"—a modest accolade that he later called a "white lie" to soothe ruffled feathers.[5] Then he goes on:

> Even before the Greenhouse test [in May 1951] it became evident to a small group of people in Los Alamos that a thermonuclear bomb might be constructed in a comparatively easy manner. To many who were not closely connected to our work, this has appeared as a particularly unexpected and ingenious development. In actual fact this too was the result of hard work and hard thought by many people. The thoughts were incomplete, but all the fruitful elements were present, and it was clearly a question of only a short period until the ideas and suggestions were to crystallize into something concrete and provable.

Oh, Edward, your human frailty is so much on display. Despite the rich scattering of names in the rest of this "many people" article (more than forty, including mine), no names appear in this paragraph.

One of the people who regarded the Teller-Ulam idea as unexpected and ingenious was the eminent theoretical physicist Hans Bethe, who called this breakthrough "an entirely unexpected departure from the previous development."[6] Bethe said also, "In January 1951, Teller obviously did not know how to save the thermonuclear program."[7] Although not employed full-time at Los Alamos in early 1951, Bethe was no doubt fully

---

*Teller was singled out for credit especially by James Shepley and Clay Blair, Jr. in their book *The Hydrogen Bomb: The Men, the Menace, the Mechanism*[2] (a book now largely discredited for its exaggerations and inaccuracies[3]).

informed. He had headed the Theoretical Division at Los Alamos during the war years and remained a valued consultant. He was known for the care with which he chose his words.

A few years later, in his book *The Legacy of Hiroshima*, Teller presents a similar account, with Ulam still factored out and Freddie de Hoffmann factored in:[8]

> I approached the problem by attempting to free myself entirely from our original concept. That done, it soon became obvious that the job could be done in other ways. During the urgent computations for Greenhouse [scheduled for May 1951], many of the hard-working physicists had participated in offhand discussions about the bomb's final design. Some of these ideas were fantastic. Some were practical. None were fully examined. They had been shoved aside by the vital need to complete the calculations for the test. With the theoretical work on Greenhouse finished, these weapons ideas could be examined in detail. Eager and anxious to come to grips with the real problem, our group at Los Alamos devoted its full attention to ways of constructing an actual bomb.
>
> About February 1, 1951, I suggested a possible approach to the problem. Frederic de Hoffmann, acting on the suggestion, made a fine calculation and projection of the idea.

This was a little too much for Stan Ulam. Although he made no public protestation, he wrote a letter to Glenn Seaborg, Chairman of the Atomic Energy Commission, on March 16, 1962, objecting to Teller's version of events.[9]

> ... the history of the new "idea" leading to the present class of designs is as follows: One day early in January in 1951 [in fact almost surely February] it occurred to me that one should employ an implosion of the main body of the device and thus obtain very high compressions of the thermonuclear part, which then might be made to give a considerable energy yield. I mentioned this pos-

sibility, with a sketch of a scheme how to construct it, to Dr. Bradbury one morning. The next day I communicated it to Edward, who by that time was convinced that the old scheme [the classical Super] might not work. For the first half an hour or so during our conversation, he did not want to accept this new possibility but in the subsequent discussion, he took to it very eagerly. In subsequent discussions we have devised certain arrangements which appear in a report written jointly by us.

It is therefore perhaps with some surprise that I noticed in the above-mentioned book [*The Legacy of Hiroshima*] a presentation of this little history stating something as follows: "I have communicated my idea to deHoffmann [sic] who calculated ... etc". In fact the joint report by deHoffmann and Teller [written by de Hoffmann and, at his initiative, signed only by Teller] is subsequent to the report 1225 by myself and Edward and it mentions explicitly the origin of the two-stage scheme and makes use of this previous report.

Frederic (Freddie) de Hoffmann, early 1960s, when he was president of General Atomics. *Courtesy of General Atomics.*

In his later published version of these events, Ulam is more circumspect. He was apparently worried about the possibility of inadvertently revealing classified information. Here is some of what he had to say: [10]

> Perhaps a change [in the outlook for the Super] came with a proposal I contributed. I thought of a way to modify the whole approach by injecting a repetition of certain arrangements. Unfortunately, the idea or set of ideas involved is still classified and cannot be described here.
>
> ...
>
> Shortly after responding [to Associate lab director Darol Froman in late January 1951] I thought of an iterative scheme. After I put my thoughts in order and made a semi-concrete sketch, I went to Carson Mark to discuss it... The same afternoon, I went to see Norris Bradbury* and mentioned this scheme. He quickly grasped its possibilities and at once showed great interest in pursuing it. The next morning, I spoke to Teller. I don't think he had any real animosity toward me for the negative results of the work with Everett† so damaging to his plans, but our relationship seemed definitely strained. At once, Edward took up my suggestions, hesitantly at first, but enthusiastically after a few hours. He had seen not only the novel elements, but had found a parallel version, an alternative to what I had said perhaps more convenient and generalized... In the following days I saw Edward several times. We discussed the problem for about half an hour each time. I wrote a first sketch of the proposal. Teller made

---

*Bradbury, the director of the laboratory, was informal and approachable. Ulam, or just about anyone else, could drop in on him with little fanfare and little advance notice. Bradbury, like Mark, understood in detail the work that was going on at the lab.

---

†Cornelius Everett, the other half of Ulam's two-person group in the Theoretical Division, was as retiring as Ulam was gregarious. When going from one place to another in the lab, Everett had the habit of walking close to the side of a corridor, tapping the wall with one hand to guide his progress, so that his thoughts could be elsewhere. He was known for the meticulous accuracy of his work. He and Ulam carried out the calculations that cast doubt on the feasibility of the classical Super.

some changes and additions, and we wrote a joint report quickly. It contained the first engineering sketches of the new possibilities of starting thermonuclear explosions. We wrote about two parallel schemes based on these principles. The report became the fundamental basis for the design of the first successful thermonuclear reactions and the test in the Pacific called "Mike." ... A more detailed follow-up report was written by Teller and de Hoffmann.

Norris Bradbury, 1950. *Los Alamos National Laboratory, courtesy of AIP Emilio Segrè Visual Archives.*

Stan Ulam's wife, Francoise, was more forthright in her discussion of the Ulam-Teller idea. Here is an excerpt from an epilogue to Stan's memoirs, which she wrote after his death: [11]

The technical and political debates were raging when Stan, mulling over the problems, suddenly came upon a totally new and intriguing approach. Engraved on my memory is the day when I found him at noon* staring intensely out of a window in our living room with a very strange expression on his face. Peering unseeing into the garden, he said: "I found a way to make it work." "What

*At that time, it was a short walk from Stan's office to his home. On the day that Francoise remembers, he had no doubt walked home for lunch.

work?" I asked. "The Super," he replied. "It is a totally different scheme, and it will change the course of history."

I, who had rejoiced that the "Super" had not seemed feasible, was appalled by this news, and anxiously asked what he intended to do. He replied that he "would have to tell Edward." Fearing that Teller might pounce on him again, I ventured that maybe he ought to test his idea on Mark or Bradbury first. He did, but went to Teller the next day just the same.

As time passed, Teller's claims about the timing of the invention of radiation implosion—and his personal role in it—became less credible, and the "work of many people" became largely the work of one person. Here is some of what he had to say in a 1979 interview with Jay Keyworth.* [12]

On January 15, 1951, which happened to be my birthday, when that meeting [with Bradbury and others] ended with agreement on how to do the testing [at Greenhouse] I said to Bradbury and now I want to go ahead with the next one, and Bradbury said it was too late, there is no point until after the test. I found this rather outrageous. The more so because by that time I knew how to solve the problem. I cannot tell you when I knew it, I think it was in December 1950. It was not long before that January meeting and I also know very accurately why I did not understand all this much sooner.

...

All this was clear to me. I don't know whether all of it was before the 15th of January. I believe so. And I believe that very shortly after that event I also told the new story to Johnny von Neumann.

If Teller had these insights before mid-February 1951, he failed to share them with those of us who were working with him on thermonuclear weapons at Los Alamos. In his 2001

*Keyworth is a physicist who served in the early 1980s as science advisor to President Reagan.

Memoirs, he moves the date of discovery back definitively to a time before Christmas 1950.[13]

> That was my state of mind on an afternoon in late November or early December [1950] when Carson Mark stuck his head into my office [to inform Teller about a visiting Admiral who responded to information about the dim outlook for the Super by saying, in effect, "Damn the torpedoes; full speed ahead"] ... Carson Mark made it a practice to needle me in a subtle manner ... the flood of adrenaline he engendered in me had real repercussions.
>
> I began to review every idea that had gone into planning for a hydrogen bomb, looking furiously for a mistake or a new idea.
>
> ...
>
> Within an hour of Carson's derisive remarks, I knew how to move ahead—avoiding the torpedoes. Thus, almost at once, the new plan appeared to be ready.
>
> ...
>
> I first went to see Johnny von Neumann, who agreed that the equilibrium approach seemed very promising. I also went over the plan with Freddie de Hoffmann, who was enthusiastic. But Darol Froman, who was then head of the Family Committee, would not really listen.
>
> ...
>
> The impossibility of a serious general discussion created a sense of unreality. That feeling grew even stronger when, around Christmas, I went to an American Physical Society meeting in Pasadena [actually in Los Angeles at UCLA].
>
> ...
>
> [After a talk by Lee DuBridge, president of Caltech and former president of the American Physical Society] I went up to DuBridge and tried to explain to him that some interesting and important work was going on at Los Alamos. He would not listen.

John Wheeler, John Toll, and I were with Teller at that December 1950 meeting. Had he shared with us at that time any thoughts on the equilibrium Super, I believe I would remember.

In his *Memoirs* Teller goes on to describe the meeting of January 15, 1951 in Los Alamos, in which Bradbury "refused me permission again!" to "get on with the new work." [14] He then goes on to describe his mid-February meeting with Ulam:

> Not long after my visit to Nevada [late January or early February 1951], Stan Ulam came to my office. He announced that he had an idea: Use a fission explosion to compress the deuterium, and it would burn. His suggestion was far from original: Compression had been suggested by various people on innumerable occasions in the past. But this was the first time that I did not object to it. Stan then proceeded to describe how an atomic explosive should compress several enclosures of deuterium through hydrodynamic shock. His statement excluded my realization of why compression was important, and it also included details that were impractical.
>
> I told him that I had thought of something that might work even better: It would be much more effective to compress the deuterium with the help of radiation ... But Stan was not interested in my proposal and refused to listen.
>
> Finally, to put an end to the discussion, I told him that I would write up both proposals, and we would sign it as a joint report. I have no idea whether Stan ever considered the extent to which compression would or would not help. But, having considered it so many times in the past, I never imagined that our joint report would be the first to discuss seriously the possibility of compression.
>
> In that paper, I wrote down my new plan for the first time. I explained how it would work and why it was better to compress the deuterium through radiation.

On February 24, 1995, I interviewed Carson Mark at his home in Los Alamos.[15] He was 81 at the time, and not in the best of health (he died two years later) but he was as feisty and as free with his opinions as I remembered him. When I asked him to comment on the "famous Teller-Ulam idea" without revealing anything that was still secret, he snapped back, "I'll say some things which I don't especially want you to recover from your damn tape," and then went on (indeed without revealing any secrets):

> [Stan's] contribution was great, very significant, but in a manner fortuitous ...
>
> He came into my office one afternoon, and I didn't really want to see him, because we were running an operation in Nevada right at that time ... All steamed up, wanting to talk about some thinking of his recently, on which he put the tag. "It would be wonderful to see the effect of a bomb in a box." He used that phrase a number of times ... Stan didn't know anything about what we were doing in that department [Nevada], came in and dismissed the whole thing, and said that it'd be a lot more interesting to explore the effects you could achieve with a bomb in a box. Put a bomb here and wrap a heavy case around it. Energy will come out and down here in the corner we can put a very small amount of plutonium and compress it to a very high degree and have a reaction from an amount of plutonium that we'd never previously thought we could do anything with. All very correct statements, not having any relation to the super at all ... He gave me this long lecture. Then the next day he went and spent an hour with Edward in Edward's office, and Edward immediately applied the ideas he was talking about to the effect they might have on a thermonuclear assembly. Instead of squeezing a little plutonium very hard and getting a reaction from a smaller amount than we'd ever considered before, Edward moved that over to taking a mass of lithium deuteride* or deuteri-

---

*Lithium deuteride, as I mentioned in Chapter 1, is the compound lithium hydride in which the hydrogen is "heavy hydrogen"—i.e., deuterium. And the lithium, although Mark does not specifically mention it, is enriched in the isotope lithium-6, which makes up only about eight percent of ordinary lithium.

um and squeezing it to a degree that we'd never imagined before. So the bomb in the box was immediately translated into an approach to a thermonuclear reaction, and Edward translated what I'd heard from Stan to a means of handling appreciable amounts of thermonuclear fuel.

When I put it all together, and despite Carson Mark's version of events, I have to conclude that Ulam did have a thermonuclear weapon in mind when he barged into Mark's office. Yet he may not have understood the full benefit of compression, and he had surely not thought about radiation as the mechanism of compression. As to Teller, I have to conclude that, despite his later testimony about his thinking in December and January, his ideas about radiation implosion and the equilibrium Super had not gelled prior to his February meeting with Ulam.

Chapter 3
The Choice

When I entered Princeton's graduate program in physics in the fall of 1948, the remotest thing from my thoughts was that two years later I might find myself in New Mexico working on the hydrogen bomb.

I was one of a dozen new graduate students in physics that fall. We all aspired to a doctorate. Most of us imagined— and indeed most of us did achieve—a career in academia. At that time, the master's degree in physics at Princeton, although it could be a stepping stone on the path to the Ph.D., served also as a consolation prize for those few who fell short of qualifying for the Ph.D. That's one reason I have no master's degree. It just wasn't something you spoke proudly of earning. The other reason was that it entailed a fee of $25, a sum I didn't want to part with in 1950 when I was offered the degree.

The standard path for physics graduate students at the time was two years of course work, followed by a several-day "qualifying exam"—mostly written but partly oral. Taking that exam had three possible outcomes: failure, a rarity; a squeaky pass, good for a consolation master's degree; and a solid pass, which opened the turnstile to dissertation research and a possible Ph.D. A couple of the group managed to get through that turnstile after only one year. I, like most of the rest of us, did so in May 1950, after two years. Although seeking out a dissertation adviser and embarking on research before taking the qualifying exam was a possibility, most of us (including me) waited. In due course, we earned our Ph.D. degrees, four in 1952 after four years as graduate students, three (including me) in 1953, and the others in 1954 and 1955, except for my brilliant, multi-talented friend Gene Saletan, who finally pulled it off in 1962. (Gene was, among other things, an accomplished folk dancer. That's an avocation I pursued for many years, although never with his special grace.) One of our group, Tor Staver, from Norway, tragically died on March 1, 1952 [1] when he lost control on a ski slope in New Hampshire and ran into a tree at high speed. The authorities at Princeton decided that he was far enough along in his research to merit the doctoral degree, and it was awarded posthumously that June. [2]

One of the students finishing in 1952 was Silvan (Sam) Schweber. Not many weeks had passed in the fall of 1948 before we had him pegged as the smartest guy (we were all male) in the group. He didn't disappoint. Sam went on to do notable work in theoretical physics, co-authoring an advanced text with the future Nobelist Hans Bethe,[3] then in his later years becoming a distinguished, prize-winning historian of science.[4] In 2012, at the age of 84, he published a biography of Bethe [5] that garnered what, in the world of scholarship, could be called rave reviews.[6]

Two other 1952 Ph.D.'s from my group were David Carter, a Canadian, and Lawrence (Larry) Wilets, from Wisconsin. Besides being fast-track graduate students, they were, as it happened, both bridegrooms at an early age. Larry was already married when he arrived in 1948, and David got married the next year. Both, as it turned out, joined the H-bomb design effort in 1951 when a branch of it moved from Los Alamos to Princeton (much more on that later). For them as for most of us who were involved, more than patriotism no doubt played a role. The pay was good (close to that of an assistant professor*) and the physics was fascinating.

In the spring semester of my first year at Princeton, I took a course on classical mechanics from John Wheeler. That is a traditionally unexciting course, just something that has to

---

*About $5,000 per year.

be absorbed and endured. Classical mechanics is the seventeenth-century science of Isaac Newton as refined and mathematically polished in the eighteenth and early nineteenth centuries. It includes none of the exciting physics of the late nineteenth and early twentieth centuries—thermodynamics, electromagnetism, relativity, and quantum physics. But, as it turned out, John Wheeler didn't believe in unexciting physics. He was determined to look at classical mechanics in a new way—for instance, to find every mathematical link it had to quantum mechanics, and to search out applications that may not have been considered before, such as to the motion of a hypothetical magnetic monopole (no real ones have been found yet) under the influence of electrically charged particles.

John Wheeler, unknown date.
Courtesy of AIP Emilio Segrè Visual Archives.

Some Princeton professors came to class meticulously prepared, delivering polished lectures and neatly filling the blackboard with equations, left to right and top to bottom. John Wheeler walked into the classroom apparently prepared only with attitude and spirit. He would say to us, in effect, "Oh, yes, where were we? What exciting thing can we explore today?" Then he would lead us down some byway that might or might not arrive at an interesting result, all the while putting on the board as many drawings as equations. (Wheeler was ambidextrous, writing, drawing, and erasing, sometimes simultaneously, with whichever hands were handy.) Some of my fellow students were drawn to the more polished professors. I was drawn to the slightly wilder Wheeler. "He's the guy I would like to work with on my doctoral research," I said to myself. A year later, I had the chance to approach Wheeler to put that thought into the form of a request.

Wheeler departed with his wife and three children for France aboard the SS *Marine Tiger* on June 29, 1949, [7] not long after completing that course on classical mechanics. Their first destination was St. Jean de Luz in the south of France, where a Princeton graduate student, John Toll, joined them. (Toll, later an important part of the H-bomb design team, and still later President of the University of Maryland system, had started his graduate work earlier than I and, in 1949, was already embarked on his dissertation research.) Wheeler's "holiday" in St. Jean de Luz naturally included physics.

Supported by the John Guggenheim Foundation, Wheeler planned to spend most of his sabbatical year 1949-1950 in Paris, "with side trips to Copenhagen," [8] where his former mentor and idol Niels Bohr held forth. As Wheeler explained in his autobiography: [9] "Although I wanted to work with Bohr, I did not want to get back fully into the conversational culture of his institute. I wanted time for isolated thinking and calculating, and knew that it would be an easy matter to travel by train from Paris to Copenhagen as often as I wished during the year." As fodder for his "isolated thinking," he had in mind a world made entirely of electrons and positrons and a world in which space and time were unnecessary, or at best auxiliary, concepts. A more down-to-earth problem that he mulled in late September, sitting in a train on his way back to Paris from his first "side trip" to Copenhagen, was why certain atomic nuclei were so markedly distorted away from spherical shapes (they possessed unexpectedly large "quadrupole moments"). [10] At the time, the "liquid-droplet" model of the nucleus, predominant in the 1930s and early 1940s, was giving way to the "independent-particle" model. Physicists were coming to understand that the nucleus had not just the properties of a liquid, but also the properties of a gas. Wheeler's insight on this train ride was to realize that the two models could be combined to account for the notably ellipsoidal shapes of some nuclei. A particle within the nucleus—a proton or neutron—could whirl around like a gas molecule in such a way as to press outward on the nuclear "skin" and distort it.

At the time, Wheeler and another of his graduate students, David Hill (also from an earlier class), together with Niels Bohr, were working on a long paper devoted to what Wheeler had christened the "collective" model—a fusion of the liquid-droplet and independent-particle models*—intended to account for properties of nuclear fission as well as other nuclear properties. Wheeler offered Bohr co-authorship of the paper mostly as a courtesy. Here is the manner in which Bohr accepted co-authorship in a letter he wrote to Wheeler on July 4, 1949: [11] "The manuscript that you sent me came as a great surprise but, realizing that it more represents an account of the discussions we through the years have had about the theme rather than some original contribution of which I feel innocent, I do not only agree with the plan, but welcome it as a token of the continuation of our co-operation." (Bohr's reputation for indirection is well deserved.) By the time the paper was finally published in 1953, after various iterations, and after various delays occasioned by Bohr's wish to review and delve more deeply, it was missing Bohr as a co-author. He

---

*The same combined model was called the "unified" model by Aage [OH-uh] Bohr (Niels Bohr's son) and Aage's colleague Ben Mottelson (an American resident in Copenhagen). Although Wheeler is famous for many coinages, including black hole and wormhole, in this case it was the Bohr-Mottelson name, not the Wheeler name, that gained currency.

had asked to be excused, since the work that was reported was indeed that of Wheeler and Hill. In the meantime, Wheeler got scooped on his explanation of nuclear deformation. In 1950, while Bohr dithered, James Rainwater, a Columbia University physicist, reported the same idea, [12] which earned him a Nobel Prize in 1975.* At the time of that prize, Rainwater wrote that the idea had come to him in late 1949. [13]

Remarkably, Wheeler never exhibited the least hostility toward Niels Bohr over this incident, saying merely: "Insights have a way of surfacing in different places at the same time." [14] "I have often wondered," he wrote in his autobiography, "whether Bohr ... let fall some remark to his son, which, carried to Columbia, was sufficient to germinate the same idea about nuclear deformation there." With a notable generosity of spirit, Wheeler went on: "It is equally likely that the flow was in the other direction. Perhaps, during my September visit, Bohr made some remark to me, based on his discussions with his son, that was just sufficient to set my mind working in the new direction on the train trip to Paris." [15]

While Wheeler was happily engaged in thinking about electrons, positrons, atomic nuclei, and spacetime, and his wife was soaking up French culture, and his children were well embarked on learning a new language, the Soviet Union exploded its first atomic bomb. The explosion occurred on August 29, 1949, and was announced publicly by President

---

*A prize he shared with Aage Bohr and Ben Mottelson, who did related work on the structure of atomic nuclei. My own early research followed lines laid out by Bohr and Mottelson. As it happened, Mottelson and I had met in Washington, DC in 1944 when we were both seventeen and participants in the Westinghouse Science Talent Search. Mottelson is one of seven alumni of that competition (now sponsored by Intel) to have won a Nobel Prize, four of the seven in physics.

Truman on September 23. This event exacerbated an already prevalent anticommunism in America, propelled the excesses of the House Un-American Activities Committee, undergirded the "McCarthy era" that was about to be launched by the junior Senator from Wisconsin, and fed a general fear that World War III might be around the corner. It markedly affected the attitude of John Wheeler, and also provided a rationale for Edward Teller to step up his demand that the Atomic Energy Commission and the Los Alamos Lab give higher priority to the development of the H bomb.

Some time that fall, probably in October, an overseas call came in for Wheeler on the wall phone at Pension Domecq in Paris, where he and his family were living. It was from Henry (Harry) Smyth, who had been Wheeler's department chair at Princeton and was now an Atomic Energy Commissioner in Washington.* [16] The call came just at dinner time. While his fellow residents watched and listened, Wheeler tried to discuss A bombs and H bombs without mentioning either. Smyth, at his end, was being similarly discreet, even without an audience. His message: Would Wheeler consider cutting short his leave in order to go to Los Alamos to participate in an accelerated effort to develop a hydrogen bomb? Wheeler didn't say no. He said maybe. [17] It's not much of a stretch to assume that this call came at the instigation of Edward Teller. Teller, even before the news of the Soviet A bomb, had taken a leave of absence from the University of Chicago to spend the 1949-1950 year at Los Alamos (a year that was later extended). The news of "Joe 1" (as the first Soviet A bomb came to be known) no doubt amplified his zeal to intensify the U. S. effort to develop an H bomb. Without question, he would have considered

---

*Smyth is best known as the author of "Atomic Energy for Military Purposes," popularly known as the "Smyth Report." It was issued in August 1945, just a few days after nuclear weapons were dropped on Japanese cities, and was intended to provide an unclassified account of the development of those weapons.

Wheeler as an ideal colleague to join the effort (as well as other notable physicists such as Enrico Fermi and Hans Bethe).

Teller directly contacted Wheeler by telegram on January 11, 1950. I have to guess that this was to put on paper one or more previous telephoned invitations. The telegram read:

WOULD YOU CONSIDER COMING TO LOS ALAMOS IN THE IMMEDIATE FUTURE AND STAY AT LEAST UNTIL OCTOBER FIRST IF POSSIBLE LONGER YOU ARE URGENTLY NEEDED WIRE COLLECT WHAT CONSIDERATIONS WOULD INDUCE YOU TO COME. [18]

(Whether Teller was authorized to offer employment at the lab is questionable, but hardly important. He could be sure that lab officials would back him up.) Wheeler, in his autobiography, describes his "inner struggle," saying that he agonized over the decision with his wife Janette, and was so "visibly troubled" that his children later remembered it. [19] Teller, in his memoirs, doesn't mention the telegram or any phone calls that fall or winter, but describes calling Wheeler some time after January 31, 1950. [20] On that date, President Truman had issued a statement saying "I have directed the Atomic Energy Commission to continue its work on all forms of atomic weapons, including the so called hydrogen or superbomb"* [21]—a statement that was later interpreted to be Truman's authorization of a "crash program" to build an H bomb. [22] According to Teller, Wheeler responded to the entreaty by saying, "Here you are supporting one end of the project and President Truman is supporting the other end, but there is nobody supporting the clothesline in the middle. I had better take the next plane." That does, in fact, sound like the kind of thing Wheeler might say, and is consistent with Wheeler's own recollection that by late January 1950 (*before*

---

*In the same statement, Truman expressed the hope that "a satisfactory plan for international control of atomic energy [will be] achieved." The statement is quoted in full in a footnote on page 41.

Truman's statement about the U.S. weapons program) he had made up his mind. In February he did depart for America, but by ship, not plane. [23]

As Wheeler recalled it, Niels Bohr gave him the final impetus to join the H-bomb work. In January 1950 Wheeler made the second visit of his sabbatical year to Copenhagen (the first having been in September), and there discussed his ambivalence. Over breakfast one morning, according to Wheeler, Bohr said, "Do you imagine for one moment that Europe would now be free of Soviet control if it were not for the Western atomic bomb?" [24] Wheeler's mind was made up. When he was back in Paris at the end of the month, his wife Janette concurred in his decision.

Thanks to the informality of the times, Wheeler could almost "take the next plane" despite having no formal job offer, no agreed-on salary, no information about travel reimbursement, and no security clearance for weapons work (although he must have had some level of clearance for his membership on the Reactor Safeguards Committee). He and his wife decided that she and the children should remain in Paris until the end of the school year (they actually cut it shorter), and that he should go ahead alone as soon as possible to the mesas of New Mexico. He encouraged his student John Toll, who could pursue his doctoral research as well in Paris as anywhere else, to stay behind with Janette and the children, which Toll gladly did. Toll had already traveled with Janette and the children as they drove in September from St Jean de Luz to Paris, and was treated as a member of the family.

John and Janette decided on a short vacation before he left. They climbed into their Renault Quatre Chevaux (a step up from the then-popular Deux Chevaux) and drove to northern Italy, with a stop in Nice to meet a remarkable young man from Los Alamos, Frederic (Freddie) de Hoffmann. De Hoffmann, born in Vienna, was twenty-five at the time and already in possession of a Harvard Ph.D. in physics. He was then the

John Toll, c. 1955. *Photograph by Al Danegger, courtesy of AIP Emilio Segrè Visual Archives, Physics Today Collection.*

chief scientific assistant to Edward Teller* and had no doubt been dispatched by Teller to brief Wheeler on the state of research on the Super. (It was in keeping with Freddie's panache that he arranged for the meeting to take place in Nice's Hotel Negresco, which he described as his favorite hotel in the south of France. [25]) When Carson Mark learned for the first time of this meeting during my interview with him in 1995, he was at first dismissive of de Hoffmann, saying that he "ran errands for Edward right from the beginning [of his work at Los Alamos]." Then Mark, becoming more animated, said: "[Y]ou say they met in Nice for a briefing, which was certainly a violation of security regulations. For one thing, he [Wheeler] shouldn't have been told about it at that stage; for another thing he [de Hoffmann] shouldn't have talked to him in any of the facilities available to them in Nice. He should have been discussing things like this only inside a controlled area. And only to a person whose clearance for the subject he was certain, which

---

*De Hoffman later became the president of General Atomics in La Jolla, California, and still later the president of the Salk Institute, also in La Jolla. He is pictured on page 16.

couldn't have happened." [26] But things were a bit looser in those days, and Freddie no doubt told Wheeler enough to re-affirm Wheeler's interest in joining the project.

En route from Paris to Los Alamos, Wheeler stopped in Princeton, probably in late February 1950, where, he wrote later, "I found something I had never experienced before: dissonance with my colleagues." [27] It was a difficult period for the country, and a difficult period for scientists, who became divided over questions of weapons work, of loyalty, and of anticommunism.

Wheeler came again to Princeton in early April, after collecting his family upon their arrival by ship in New York from France. [28] I took advantage of his presence in Princeton then to pop the question: Would he be willing to guide my doctoral dissertation work? "Yes, certainly," was his answer, "but you should know [I paraphrase] that I will be away from Princeton for at least a year to work in Los Alamos on the hydrogen bomb." He went on to say that he would be pleased if I decided to join him there, as his student John Toll already planned to do. He made clear that if I came, I would be expected to work at the lab on the H bomb project but that I would probably find some time for pure research as well—a division of effort that he himself planned to embrace.*

In modern parlance, Wheeler gave me a soft sell. The hard sell came a few weeks later—in late April or early May—when Teller came to town. It was my first exposure to Teller's bushy eyebrows and persuasive powers. On a lovely spring day we sat together on the steps of Fuld Hall at the Institute for Advanced Study and he explained to me the urgency of the

---

*In those informal times, Wheeler felt free to offer me a job at Los Alamos, well outside his official responsibilities.

project and the reasons that I should join it. I told him, as I had told Wheeler earlier, that I would think about it.

I consulted with a few friends and with Allen Shenstone, the physics department chair, but not with my parents, for I knew that they would endorse whatever I decided. Shenstone opposed my taking a leave of absence—not, he said, because of aversion to weapons work but because he feared that I might never come back and complete my graduate studies. He knew of other cases, he told me, in which a graduate student on leave became absorbed in new activity and never returned to earn a Ph.D. I was not moved by that argument because I felt certain that I would indeed return and finish my graduate studies (as I did). Separately, Shenstone had let Wheeler know that he opposed Wheeler's own plan. Wheeler's research and his teaching duties, Shenstone argued, were simply more important than pursuit of a new weapon. [29]

At the time politics, like professional sports, were outside the range of my interests. Yet I had a general feeling that it would be a good thing if teams from Boston won their games and if the United States acquired an H bomb before the Soviet Union did. I thought of the United States as a moral nation that could be trusted with weapons of nearly unlimited destructive power, and the Soviet Union as a nation that could not be trusted. So, despite my naiveté and my disconnect from world affairs, I made a decision based on essentially the same arguments that led President Truman to make his decision.

There were other factors that led me to head for Los Alamos. I was drawn to new challenges and thought that working on the H bomb would be "fun." And the idea of close daily contact with John Wheeler, whom I so much admired, was appealing. As it turned out, working on the bomb *was* fun, and working closely with Wheeler was rewarding (leading to a lifetime friendship that went beyond student-professor).

I let Wheeler know that I would show up in Los Alamos in late June. At once wheels started turning that resulted, surprisingly quickly, in a formal job offer and the granting of security clearance. In the meantime, there was a qualifying exam to be taken and there were cars to be acquired and disposed of.

Chapter 4

The Scientists, the Officials, and the President

On October 28–30, 1949, just five weeks after President Truman's announcement of the Soviet atomic bomb test, the General Advisory Committee of the Atomic Energy Commission met in Washington, in part to address the question of whether to "pursue with high priority the development of the super bomb." [1] It is difficult to imagine a committee containing a higher percentage of distinguished physical scientists and scientists with links to the Manhattan Project than this one*, yet, as it turned out, the recommendations the committee reached after more than two days of intense discussion had no discernable effect on the prosecution of weapons design work.

After a full weekend that began with a Friday evening session, the committee's report was in three sections: a main report, a majority annex, and a minority annex.[3] All three sections recommended against giving priority to the development of a super bomb. The main report states: "We

---

*J Robert Oppenheimer, Chair, a physicist, Director of the Institute for Advanced Study, wartime director of the Los Alamos lab; Oliver E. Buckley, a physicist, president of Bell Labs; James B. Conant, a chemist, president of Harvard University; L. A. DuBridge, a physicist, president of Caltech; Enrico Fermi, professor of physics at the University of Chicago, Nobel Prize 1938; I. I. Rabi, professor of physics, Columbia University, Nobel Prize 1944; Hartley Rowe, an electrical engineer, vice president of United Fruit Company; Glenn Seaborg, professor of chemistry, UC Berkeley, to be awarded a Nobel Prize in 1951 (absent from this meeting); Cyril Stanley Smith, a metallurgist and historian, University of Chicago. [2] Most of these members played key roles in the Manhattan Project.

all hope that by one means or another, the development of these weapons can be avoided. We are all reluctant to see the United States take the initiative in precipitating this development. We are all agreed that it would be wrong at the present moment to commit ourselves to an all-out effort toward its development." At the same time, the main report cites, as the principal reason for its negative recommendation, "the technical nature of the super and of the work necessary to establish it as a weapon," and offers the opinion that "an imaginative and concerted attack on the problem has a better than even chance of producing the weapon within five years." It is not so surprising, then, that twenty months later (in June 1951), after the likely success of the Teller-Ulam idea was established, the General Advisory Committee reversed itself and showed enthusiasm for the rapid development of the H bomb. [4] The first thermonuclear explosion took place three years, almost to the day, after the October 1949 report.

The majority annex, signed by Conant, Rowe, Smith, DuBridge, Buckley, and Oppenheimer, cited moral rather than technical reasons for opposition to an "all-out" effort to develop the super bomb. "We recommend strongly against such action," this group wrote. "We base our recommendation on our belief that the extreme dangers to mankind inherent in the proposal wholly outweigh any military advantage that could come from this development ... [T]his is a super weapon; it is in a totally different category from an atomic bomb ... Its use would involve a decision to slaughter a vast number of civilians ... If super bombs will work at all, there is no inherent limit in the destructive power that may be attained with them. Therefore, a super bomb might become a weapon of genocide." And, finally: "In determining not to proceed to develop the super bomb, we see a unique opportunity of providing by example some limitations on the totality of war and thus of limiting the fear and arousing the hopes of mankind."

J. Robert Oppenheimer, unknown date. *Digital Photo Archive, Department of Energy (DOE), courtesy of AIP Emilio Segrè Visual Archives.*

The minority annex, signed by Fermi and Rabi, went even further in its abhorrence of the very idea of a "Super." "Necessarily," they wrote, "such a weapon goes far beyond any military objective and enters the range of very great natural catastrophes." And, further, "It is clear that the use of such a weapon cannot be justified on any ethical ground which gives a human being a certain individuality and dignity even if he happens to be a resident of an enemy country." They added this now-famous comment, "It is necessarily an evil thing considered in any light," and went on to recommend that the President "tell the American public, and the world, that we think it wrong on fundamental ethical principles to initiate a program of development of such a weapon." They went on: "At the same time it would be appropriate to invite the nations of the world to join us in a solemn pledge not to proceed in the development or construction of weapons of this category."

Since every member of the General Advisory Committee who took part in these deliberations was a party to either the majority or the minority annex and thus expressed opposition to the Super on moral grounds, it is a bit odd that the commit-

Enrico Fermi in Los Alamos, c. 1945.
*Photo by Sgt. E.D. Wallis, courtesy
of AIP Emilio Segrè Visual Archives,
Bainbridge Collection*

tee as a whole, in its main report, cited technical difficulty as the principal reason for opposing the development. It is likely that among all the members, only Oppenheimer and Fermi had a deep enough understanding of the underlying physics to have reached an informed opinion about the level of technical difficulty in making a Super. The other members no doubt accepted the testimony of these two in endorsing the conclusions stated in the main report. As it happened, Fermi pitched in and became a major contributor to the H-bomb development after President Truman authorized it in late January 1950. (It was my privilege to work with Fermi at Los Alamos in 1950-51.) Oppenheimer, although he made no personal contributions to the development, switched from opposition to support once he became convinced, in June 1951, that, with the Teller-Ulam idea, the H bomb was quite feasible. (More later on the meeting that probably changed his mind.)

The Commission forwarded the advisory committee report to President Truman on November 11, 1949, a little less than two weeks after the report was completed, with what amounted to a lukewarm endorsement. [5] Of the Commission's five members,* just three supported the report's recommendations. The supporters were the Commission Chair, David Lilienthal; Sumner Pike; and Harry Smyth.† The dissenters were Gordon Dean and Lewis Strauss. Strauss, a confidant of Edward Teller, had been pushing hard since the news of the Soviet atomic bomb test in September for an accelerated program to develop a thermonuclear weapon. He took his case separately to the President in a letter dated November 25, 1949, a letter in which he mentions that his views are supported by his fellow Commissioner Gordon Dean. [7] Dean was apparently aligned with Senator Brien McMahon on nuclear weapons issues. McMahon, who had authored the 1946 bill that created the Atomic Energy Commission and was "Mr. Atomic Energy" in the Congress, vigorously advocated a priority effort to develop an H bomb. [8]

In retrospect, it is easy to see that the go-ahead provided by Truman on January 31, 1950, was inevitable.‡ He had a bare

---

*The Atomic Energy Commissioners as of fall 1949 and early 1950: David E. Lilienthal, Chair, a lawyer and former head of the Tennessee Valley Authority—also a dedicated diarist; Gordon Dean, a lawyer who had worked in the Justice Department and in the law firm of Brien McMahon (McMahon, at this time, was a member of the U.S. Senate and chaired the Joint Committee on Atomic Energy); Sumner Pike, a businessman who had been a member of the Securities and Exchange Commission; Henry DeWolf Smyth, professor of physics at Princeton University; and Lewis Strauss, a banker and admiral in the U.S. Navy Reserve. [6]

---

†Smyth placed his call to Wheeler in Paris before the General Advisory Committee report. Had Smyth known what the GAC would recommend, it seems unlikely that he would have sought to recruit Wheeler.

---

‡Here is the full text of Truman's January 31, 1950 statement: "It is part of my responsibility as Commander in Chief of the Armed Forces to see to it that our country is able to defend itself against any possible aggressor. Accordingly, I have directed the Atomic Energy Commission to continue its work on all forms of atomic weapons, including the so called hydrogen or superbomb. Like all other work in the field of atomic weapons, it is being and will be carried forward on a basis consistent with the overall objectives of our program for peace and security.

"This we shall continue to do until a satisfactory plan for international control of atomic energy is achieved. We shall also continue to examine all those factors that affect our program for peace and this country's security." [9]

majority of the Atomic Energy Commission supporting the advisory committee's go-slow recommendation. He had influential advice favoring full speed ahead not only from Lewis Strauss but from Louis Johnson, the Secretary of Defense, from the generals in uniform, [10] and from Senator McMahon. [11] At the same time, he could hardly have avoided absorbing what was becoming public advocacy by McMahon and like-minded politicians. [12] And, just beginning to sweep the country, as the Cold War heated up, was an anticommunist fervor that made it hard to contemplate any step that might entail trust in, or cooperation with, the Soviets. No argument on moral grounds, it was widely believed, could have any meaning in dealing with what Strauss called the "atheists" of the Soviet Union. [7]

In November, after receiving the Commission's tepid endorsement of the advisory committee's report, Truman appointed a three-man committee to advise him on the matter. Its members were Louis Johnson, in favor of full steam ahead on the H bomb; David Lilienthal, already on the record in opposition to a priority program; and Dean Acheson, the Secretary of State, in principle neutral. [12] Acheson was Truman's closest, most trusted advisor [13] and in fact leaned toward pursuing the H-bomb program without delay. Acheson was not a knee-jerk hawk but had reached the conclusion that the United States should have at its disposal any and all weapons that were feasible to design and build. [14] By the time this committee walked into the President's office with its recommendation and its proposed statement on January 31, the President very likely already knew what he was going to do, [15] and was probably pretty sure that the committee, by at least two to one, was going to support the position he had reached. In fact, Lilienthal had come around to the belief that the country should go forward with an all-out effort, so the committee's recommendation was unanimous. The President read and approved the proposed statement and it was promptly handed out to waiting reporters, who rushed for the phones. By some accounts, this meeting lasted only seven minutes. [16]

Edward Teller, in his memoirs, expresses his great apprehension at what he feared might be the General Advisory Committee's recommendation and mentions a good meeting he had in November 1949 with Senator McMahon in Washington. [17] It was on a swing to the east in which he also called on Fermi in Chicago and Bethe in Ithaca in an unsuccessful effort to get one or both to drop their academic work and join him in Los Alamos. The day after Teller and Bethe met in Ithaca, the two went on to Princeton to see Oppenheimer, who had invited Bethe and who then enlarged the invitation to include Teller when he learned that the two were together. In his memoirs, Teller recalled that at the Princeton meeting, Oppenheimer, characteristically, kept his personal opinions to himself, although he shared with his visitors a letter from James Conant that contained a strongly worded condemnation of an H-bomb program. [18] Bethe, according to Teller, both in Ithaca and in Princeton, committed himself to joining Teller in Los Alamos, but less than a week later changed his mind. [19]

On this trip, neither Fermi nor Oppenheimer shared with Teller the content of the General Advisory Committee report in which they had participated, although Fermi was more than likely open in discussing his personal views. As for McMahon, he, according to Teller, said "Have you heard about the GAC [General Advisory Committee] report? It just makes me sick." [20] A few weeks later, back in Los Alamos, Teller was allowed to read the GAC report, which confirmed his worst fears. [21] Then, on January 31, the President's statement gave Teller a lift. In his memoirs, Teller wonders what brought Truman to the right way of thinking. "Was [Truman's] decision based simply on his abundant common sense? Probably no one will ever know [what convinced the President]," Teller continues, "but my bet is on the common sense." [22]

Chapter 5

Nuclear Energy

When Henri Becquerel, in Paris, discovered radioactivity in 1896, [1] my parents were pre-schoolers. I mention this fact only to emphasize that the history of nuclear energy from Becquerel to bombs, from a few relatively harmless alpha, beta, and gamma rays to the destruction of cities and the obliteration of a Pacific island was accomplished in one human lifetime. In 1952, the year in which "Mike" released its ten megatons and Elugelab was no more, my parents turned sixty.

What Becquerel discovered was that a uranium compound emitted some kind of "radiation" that could darken a photographic plate, even if the compound was not "activated" by shining light upon it or stimulated in any other way. The compound, wrapped in paper and kept in a dark drawer, continued, with no apparent diminution of intensity, to emit its radiation. He also found that uranium metal alone had the same property and that seemingly no other element did.

Henri Becquerel. *Photograph by Gen. Stab. Lit. Anst. (Generalstabens Litografiska Anstalt), courtesy of AIP Emilio Segrè Visual Archives, William G. Meyers Collection.*

Becquerel had no idea that he was dealing with nuclear energy. He was probably not even sure that atoms existed, much less that atoms—if they did exist—might have tiny nuclear cores at their centers, or that the radiation he discovered might come from such cores. But he did infer that uranium must contain *stored* energy—energy that could leak out over time, and a lot of it, since it did not weaken over the days and months that he studied it. [2]

Becquerel's "rays" drew less scientific attention at the time than the recently discovered X rays, with their seemingly magical property of revealing a person's bone structure. Wilhelm Röntgen had announced his discovery of X rays on New Year's Day 1896.* [3] Becquerel's first report on his new penetrating radiation came less than two months later, on February 24, 1896. [4] After that, a year and a half elapsed before a thirty-year-old doctoral candidate at the Sorbonne in Paris, Marie Curie, chose to follow up Becquerel's work for her dissertation research. She wanted a topic that would give her time to get new results without undue risk that some other researcher would preempt her findings. [5] Uranium, she said, is "radio-actif," and the name stuck. [6]

Marie Curie. *Courtesy of AIP Emilio Segrè Visual Archives, W. F. Meggers Collection.*

*Röntgen, a German, chose to announce his discovery in Austria.

As it turned out, Marie Curie opened a floodgate. Within a year, she and her husband Pierre had discovered two new elements, polonium and radium. Soon thereafter Ernest Rutherford, at McGill University in Montreal, discovered radioactive substances with shorter half lives, one of one minute and another of eleven hours, and he verified that their decay followed a simple probabilistic rule. In 1898, Rutherford named the two then-known kinds of radioactive emissions alpha and beta rays, and he later added the coinage gamma rays for a third kind of radiation discovered in 1900 by Paul Villard in Paris. [7]

By 1904, in his 382-page tome *Radio-Activity* [8] (the hyphen was soon dropped), Rutherford could report the following conclusions, a mind-filling set of ideas unknown and unsuspected less than a decade earlier (these are paraphrases).

- Radioactivity supports the idea that atoms exist, and suggests that they are complex structures.
- Radioactivity is a series of spontaneous explosive changes in atoms; it is not a process of gradual change.
- Radioactivity transmutes one element into another, which no chemical change can do, and has produced hitherto unknown elements.
- In radioactivity, the energy released per atom is enormous, at least a million times greater than in chemical change.
- The intensity of radiation from a given radioactive element diminishes according to a law of exponential change with a characteristic half life, suggesting that a law of probability operates at the individual atomic scale.
- Helium is emitted in radioactive decay, and alpha particles are probably helium atoms (they were later confirmed to be helium nuclei).
- The beta rays emitted in radioactive decay are electrons, and they shoot out with great energy.

Ernest Rutherford. *U.K. Atomic Energy Authority, courtesy of AIP Emilio Segrè Visual Archives.*

In 1905, the year after this monumental summing up by Rutherford, Albert Einstein offered the world his most famous equation, $E = mc^2$: Energy is mass times the square of the speed of light. Specifically, in the form $m = E/c^2$ the equation tells how much change of mass is required to produce a certain amount of energy. Because of the enormously large value of c^2 by normal standards, it takes only very little mass change to produce a great deal of energy. If, for instance, two hydrogen atoms join with an oxygen atom to form a water molecule—a vigorous combustion process that releases energy—the mass of the molecule is less than—but imperceptibly less than—the sum of the masses of the three atoms. Einstein recognized that in ordinary chemical change the changes of mass would be too small to measure. He then asked himself if there was any chance of verifying the correctness of the equation experimentally, and he made this suggestion: "It is not impossible that with bodies whose energy content is variable to a high degree (e.g. with radium salts) the theory may be successfully put to the test." (Original in German.) [9]

Nuclear energy was in the air! And the discovery of the nucleus was another half-dozen years in the future.

47

Rutherford, a towering figure of this period, quite naturally wanted to understand the interior of the atom, which he now assumed to be complex and to include moving electric charges, probably electrons. He knew, too, that the atom must contain enough positive charge to balance the negative charge of the electrons, but what carried the positive charge and how it might be distributed within the atom no one knew. Some other physicists at the time constructed models of what an atom might look like. [10] Rutherford did experiments. And he had atomic bullets available. The alpha particles shot out by radioactive nuclei could serve as projectiles to be fired at targets. If the target was a thin sheet of metal, most of the alpha particles fired at it emerged on the other side, with varying small deflections away from their original flight direction. This was no surprise. The alpha particles passed through or quite near many atoms in the thin sheet, and Rutherford assumed that at each encounter the alpha particle suffered some small deflection, which could add to or subtract from a previous deflection. A large total deflection was not expected because of the random nature of the individual deflections. It's as if you threw baseballs one after another through a stand of wheat. At each encounter with a stalk of wheat, a baseball would suffer a tiny deflection—left, right, up, down. If the stand were thin enough to allow most of the baseballs to get through, they would fan out on the other side, but only through a small range of angles. For one ball to get "reflected" and come back toward the thrower would require an incredibly improbable series of deflections, one after the other, all bending the trajectory in the same way to produce one large deflection.

This was the situation with alpha particles and metal foils in Rutherford's Manchester laboratory in 1908-1909. (Rutherford had moved in 1907 from Montreal to Manchester, UK. [11] His 1908 Nobel Prize in Chemistry did not seem to slow him down at all.) Beginning in 1909, Rutherford's associates Hans Geiger (yes, of the Geiger counter) and Ernest Marsden were beginning to see some larger-than-expected angles of deflections of alpha particles passing through gold foils. [12] This understandably caught Rutherford's attention, for it was unexpected. He initiated a series of alpha-particle scattering experiments that culminated in 1911 with his announced discovery of the atomic nucleus. [13]

Geiger and Marsden were able to measure just what fraction of the incoming alpha particles were deflected through various angles, from zero degrees all the way to nearly 180 degrees. From these results, analyzed mathematically, Rutherford drew two conclusions. First, the deflection of an alpha particle was the result not of many accumulating small deflections, but of a single encounter within an atom. Second, the force causing the deflection was an electric force resulting from the presence within the atom of a massive nugget of electric charge. [14, 15] From the experiments alone, it was not possible to tell whether this charge was positive or negative, but Rutherford assumed, correctly, that it was positive, given the evidence that negative electrons also existed in atoms. It was also not possible to tell how large this charged "nugget" (soon to be called a nucleus) was. From the number of alpha particles that "bounced" back, almost reversing course, Rutherford could conclude that this central nucleus was less than a thousandth the size of the atom (less than a billionth of the volume). [14]

It is in fact even a good deal smaller than that. Writing later about these findings, Rutherford said, "It was almost as incredible as if you fired a 15-inch shell at a piece of tissue paper and it came back to hit you." [15]

This discovery, described as introducing the "planetary model" of the atom, gave rise, over the next fifteen years, to a string of discoveries in atomic physics culminating in quantum mechanics and all of its wonders. Our concern here, however, is just with the nucleus itself and its energy. It was immediately evident to Rutherford and others that the atomic nucleus must be the site of radioactivity. Evident that unstable nuclei can emit alpha, beta, and gamma particles; that for alpha and beta emission, the nucleus is transmuted into that of a different element; and that the mass of a nucleus measures its energy content.

But could humankind control and harness the enormous energy stored within the nucleus, not just observe it? It was a writer, not a physicist, who first suggested that possibility. In his book *The World Set Free*, published in 1914, [16] H. G. Wells imagined an "atomic bomb," a device for which scientists had figured out a way to make radioactive elements release their stored energy much more rapidly than the normal pace of spontaneous decay in nature. His hypothesized new element, carolinium, had a half life of seventeen days instead of the 1,600-year half life of radium or the even longer half life of uranium. Its energy release was activated by an "inductive" applied just after the bomb, a two-foot-diameter sphere, was dropped by hand from an airplane. Then, for weeks, the bomb continued to spew out its great store of energy on unlucky combatants on the ground.

Nearly twenty years later, in 1933, Leo Szilard, a Hungarian physicist then in London, came up with another idea for a nuclear bomb, still hypothetical but far better grounded than H. G. Wells' fascinating fantasy. His thinking was based on two important discoveries of the preceding year. One was the discovery of the neutron by James Chadwick in Cambridge, England. [17] The neutron produced an immediate "aha" moment for physicists, who recognized at once that this neutral particle, about as massive as a proton, must be a constituent of atomic nuclei. Suddenly, nuclei could be imagined as collections of protons and neutrons rather than protons and electrons (a view that had been problematic for some time since no one could see how electrons could be confined within a nucleus). The other discovery of 1932 that influenced Szilard came from an experiment by John Cockcroft and Ernest Walton, also in Cambridge. They used an early-model accelerator to fire protons at a lithium target, and observed alpha particles emerging from the collision, each with an energy greater than the energy of an incident proton. Specifically, the isotope lithium-7 took part in the reaction, which can be written p + Li^7 → 2α + energy.* Cockcroft and Walton knew that the combined mass of a proton and a lithium-7 nucleus was greater than the mass of two alpha particles, and indeed this known mass difference appeared as the energy of motion (the kinetic energy) of the emerging alpha particles. [18]

Leo Szilard. *Courtesy of Bulletin of the Atomic Scientists and AIP Emilio Segrè Visual Archives.*

Szilard linked these findings in his mind by asking himself the question (while on a walk in London and waiting for a light to change, as he later reported [19]): What if an energy-

---

*This reaction can be viewed as a kind of fission process. The combination of a proton and a Li^7 nucleus creates, momentarily, a beryllium-8 nucleus (containing four protons and four neutrons), which then splits into two alpha particles, each with two protons and two neutrons. This fission process is an exception to the rule explained later in this chapter that fission releases energy for heavy nuclei, whereas fusion releases energy for light nuclei. That rule applies only for stable or long-lived nuclei. The highly unstable, and very short-lived Be^8 nucleus *does* release energy when it undergoes fission.

releasing nuclear reaction were triggered not by a proton, as in the Cockcroft-Walton experiment, but by a neutron, and what if, from the reaction, two neutrons emerged? These released neutrons could stimulate more reactions of the same kind, and one would have a nuclear chain reaction, potentially releasing vast energy. The idea of a chain reaction existed already in chemistry, and could potentially be explosive, but, as Szilard knew, a nuclear chain reaction, if it were to occur, might outdo the chemical chain reaction a million-fold.

Szilard did not imagine nuclear fission. That came more than five years later—and was a total surprise when it did come. He was thinking instead of a reaction like the one achieved by Cockcroft and Walton, but with the incident particle being a neutron instead of a proton, and the emitted particles being two neutrons instead of two alpha particles. To be sure, the Cockcroft-Walton experiment released nuclear energy, but the nuclear energy that came out if it was far less than the energy put into the accelerator that supplied the protons used to bombard the lithium. In a speech delivered shortly before Szilard's walk in London, Rutherford (by then Lord Rutherford), aware of this imbalance between the energy put into the machinery and the energy released in a nuclear reaction, had said "anyone who looked for a source of power in the transformation of the atoms was talking moonshine." [20] This remark, duly reported in The Times of London on September 12, 1933—the very day of Szilard's walk—bothered Szilard and contributed to his invention of the idea of a nuclear chain reaction. As he said later, "Pronouncements of experts to the effect that something cannot be done have always irritated me." [21]

A bit more entrepreneurial than most scientists, Szilard applied for and, in 1934, was granted a patent on the idea of a nuclear chain reaction, a patent that he soon assigned to the British Admiralty as a way to keep it secret. His application to conduct experiments in search of a nuclear chain reaction at Rutherford's laboratory in Cambridge was turned down, but he managed to conduct some experiments, first at St. Bartholomew's Hospital in London, and then, in late 1938, in Rochester, New York (he had just moved to the United States and, in typical Szilard fashion, was bouncing around among labs). In neither place did he find any evidence for such a reaction.* [22] It is hardly surprising that when the news of nuclear fission reached New York in January 1939, Szilard was among the first to see its possibilities for generating a chain reaction and for providing a weapon of surpassing power. [24]

The discovery of fission is an oft-told story. [25] In brief: In Berlin in 1938, the German chemists Otto Hahn and Fritz Strassmann, who had been bombarding uranium with neutrons to see if heavier elements might be formed, found evidence of the element barium being created. This was totally puzzling to them, yet, after the most careful checks and cross checks, the barium would not go away. Barium is element number 56, while uranium is number 92. The atomic weight of barium is 137, not much more than half of uranium's atomic weight of 238. Where was the barium coming from? Hahn sent off a letter to his former physicist colleague Lise Meitner to see if she might have an explanation. Meitner, a Jew, had had to flee Germany, and was then in Sweden. As it happened, her nephew Otto Frisch, also a physicist, and then working not so far away at Niels Bohr's institute in Copenhagen, came to spend the Christmas 1938 holiday with his Aunt Lise, and was there when Hahn's letter arrived. On a snowy trek through the woods, Frisch (on skis) and Meitner (keeping up on foot) pondered the matter and asked themselves: Could uranium nuclei, stimulated by neutrons, be splitting apart into smaller nuclear fragments (which could include barium nuclei)? Excited by the

---

*On December 21, 1938, after the negative results in Rochester, Szilard wrote to the British Admiralty asking that his chain-reaction patent be withdrawn. Five weeks later, on January 26, 1939, with the news of fission in hand, he sent a telegram to the Admiralty saying KINDLY DISREGARD MY RECENT LETTER. [23]

idea, they sat down on a tree trunk, pulled out some scraps of paper, and started to calculate, working from a formula that Meitner had in her head, the so-called Weizsäcker mass formula. This was a "semi-empirical" formula advanced by Carl Friedrich von Weizsäcker in 1935 [26] (and later refined) that provided the masses of nuclei to good approximation across the whole periodic table. Their conclusion: Breaking a uranium nucleus apart into two large fragments would release energy, a lot of energy. Their estimate was 200 MeV, which proved to be right on the mark. [27]

Lise Meitner. *Courtesy of AIP Emilio Segrè Visual Archives.*

Otto Frisch. *Courtesy of AIP Emilio Segrè Visual Archives, Physics Today Collection.*

Once Frisch was back in Copenhagen, he hastened to Niels Bohr to report his and Meitner's "speculations" about the breakup of the uranium nucleus.* Bohr, set to leave for America in a few days, immediately accepted the idea, exclaiming, according to Frisch, "Oh what idiots we all have been! Oh but

---

*I don't know if Frisch used the term "fission" in this meeting with Bohr. If not, he must have introduced it within a few days thereafter, for it is a term that Bohr brought with him to New York on January 16, 1939, after a nine-day crossing of the Atlantic. Frisch got the term from an American biologist visiting in Copenhagen, William Arnold, whom he asked what cell division is called. "Fission" was the answer. [28]

this is wonderful! This is just as it must be!" [29] By the thir-
teenth of January, while Bohr was en route to America, Frisch
had conducted experiments that directly confirmed the reality of nuclear fission.

On his week aboard the MS *Drottningholm*, Bohr convinced himself that indeed the process made great sense, and
he had no hesitation in reporting it as real when he reached
New York. Nevertheless, he limited his discussion of fission at
first to a few colleagues at Columbia and Princeton Universities—not from any sense of the military potential of fission,
but to give time for Meitner and Frisch to prepare a paper for
publication and to get the credit they deserved. As it happened,
the Meitner-Frisch paper was published with lightning speed.
It was submitted to the journal *Nature* on January 16, 1939, and
was published on February 11. [30] On January 26, Bohr gave a
public report at a conference in Washington, D.C., after which
the news spread quickly across the country. (Probably on January 30, the physicist Luis Alvarez came across a newspaper
report of the discovery of fission while getting his hair cut in
a Berkeley, California barber shop. He reportedly leaped from
his chair without waiting for the barber to finish, and hurried
to his lab. By the next day, he and his student Phil Abelson had
verified the reality of nuclear fission.) [31]

The mass spectrometer, invented by Francis William Aston in 1919, [32] made it possible to measure the masses of individual atoms*—at first with enough precision to clearly dis-
tinguish different isotopes of the same element, later with the
greater precision needed to establish that the total mass of
nuclei after a nuclear reaction need not be exactly the same as

---

*The mass spectrometer actually measures the mass of an ion (a charged
atom). From the ion's mass it is easy to infer the mass of the uncharged atom
and also the mass of its nucleus, since the electron's mass is known.

the total mass before. For example, Cockcroft and Walton, in their 1932 experiment, knew that the masses of the proton and lithium-7 nucleus added to more than the mass of two alpha particles. The mass difference, they found, was accounted for by the energy of the emitted alpha particles. The books were balanced, not on mass alone, but on mass-energy.

The Cockcroft-Walton experiment is often cited as the first experimental proof of Einstein's mass-energy equivalence. Actually there were hints of its correctness a dozen years earlier. (Einstein himself, we can be confident, had no doubts about it.) By 1920 there was evidence that the proton—the nucleus of the most common isotope of hydrogen—was just a tad "overweight." With the masses of the most common isotopes of carbon, nitrogen, and oxygen pegged at 12, 14, and 16 units, and other known isotopes also following very closely a whole-number rule, the mass of the lightest isotope of hydrogen was not exactly 1, it was about 1.01. [33] This slight oddity was just enough to make Arthur Eddington in England suggest that energy would be released if four hydrogen nuclei fused to make a helium nucleus (with electrons participating

Arthur Eddington. *Courtesy of AIP Emilio Segrè Visual Archives, gift of Subrahmanyan Chandrasekhar.*

to preserve charge conservation) and that such fusion might be the source of the Sun's (and other stars') energy. [34]

So nuclear fusion entered the consciousness of physicists nearly twenty years before nuclear fission did. And—unlike with the startling discovery of fission—there was nothing particularly surprising about the idea of fusion. Physicists (and chemists, and astronomers) assumed that nuclei were composed of smaller entities (initially supposed to be protons and electrons) and that these entities were held together by a "binding energy," which, in accordance with $E = mc^2$, would make the nuclear mass less than the sum of the masses of its constituents. So, just as it would take energy to pry apart a nucleus into its component parts, energy would be released if these parts came together to form a nucleus. Moreover, since the earliest days of radioactivity, it was clear that on a per-atom basis, these nuclear energies would be much greater than chemical energies.

In the years following Eddington's imaginative leap, other physicists explored the possibilities of fusion as the source of stellar energy. Following the discovery of the neutron in 1932 and the refinement of mass spectroscopy in the 1930s, it became possible to predict with some accuracy just how much energy would be released in a variety of possible fusion reactions. Finally, in 1939, just on the heels of the discovery of fission, Hans Bethe, a brilliant émigré physicist from Germany, then at Cornell University, put it all together and suggested two principal fusion cycles that might power stars, one involving principally hydrogen and helium, the other involving also carbon, nitrogen, and oxygen as intermediaries. [35] Astrophysicists continue to believe that Bethe got it right, and that his fusion cycles are the main sources of stellar energy. (Bethe was awarded the Nobel Prize in Physics in 1967.)

It's an oddity of history that in the very year that fusion was established as a reality in the cosmos and fission as a reality here on Earth, Hitler launched an attack on Poland,

and World War II was under way. Quite naturally, physicists asked themselves: Can fission and/or fusion be harnessed to produce practical power for humankind? Can one or both be exploited to make powerful weapons of warfare? Needless to say, the emphasis at the time was on the latter question.* As it turned out, controlled fission (a nuclear reactor) and explosive fission (an A bomb) were both achieved within half a dozen years. Explosive fusion (the H bomb) came seven years after the fission bomb. Controlled fusion for power production remains a yet-to-be-achieved goal.†

Hans Bethe, 1975. *Courtesy of Fermi National Accelerator Laboratory and AIP Emilio Segrè Visual Archives, Physics Today Collection.*

---

*There remains debate about what aspect of fission energy German scientists chose to emphasize—controlled energy or explosive energy. Every scientist or historian has his or her own opinion. Mine is that Heisenberg's "Uranium Club" would have vigorously pursued the explosive option if its members foresaw the possibility of success before the war ended. Instead they assumed that the war would end well before a nuclear bomb could be achieved, and accordingly focused on building a reactor, with no sustained high-priority push toward a bomb.

---

†There is a joke among fusion scientists that fusion power is a decade away and always will be.

Chapter 6

Some Physics

Here is some physics related to thermonuclear weapons, for those who want to read it.

Fission and Fusion

Both fission and fusion release nuclear energy, which, as the previous chapter has made plain, is vastly greater per atom or per unit of mass than the energy of chemical change. Despite the huge difference in scale, there is one thing that nuclear energy and chemical energy do have in common—they can be released either explosively or gradually. For chemical energy, think dynamite or gunpowder vs. a candle flame. For fission, think Hiroshima vs. that relatively benign power reactor up the river. For fusion, think H bomb vs. the yet-to-be realized fusion reactor that will use deuterium from the ocean to produce electricity. (As for the Sun: In one sense it releases energy gradually, over billions of years; but in another sense, it is a nonstop nuclear explosion, rather like the carolinium imagined by H. G. Wells.)

Humankind discovered chemical energy long ago, first fire, then gunpowder—the gradual before the explosive. For fission energy, the gradual and the explosive were more nearly simultaneous. Fission reactors—the gradual—actually came first, by a few years, but these early reactors produced no usable power. They served to establish the principle of large-scale fission energy and to produce plutonium for weapons. Then, a dozen years after the explosive release of fission energy, came power-producing reactors. In the long span of

59

history, gradual fission energy and explosive fission energy came at pretty much the same time. For fusion energy, the explosive came first. More than half a century after the first fusion explosion (that is, the H bomb), the gradual release of fusion energy remains a hope, not a reality.

Nuclear energy is all about $E = mc^2$. It is also all about the forces that exist within the nucleus. Nuclei are composed of neutrons and protons (collectively called nucleons).* Two kinds of force are at work. The nuclear force, or strong force, acts to attract neutrons to each other, protons to each other, and protons to neutrons. In short, it attracts every particle within the nucleus to every other one. The electric force acts to repel the positively charged protons from one another and has no effect on the electrically neutral neutrons. Besides the forces, there is a physical principle also at work: It is called the Pauli exclusion principle. I won't go into this principle except to say that its effect is to favor an equal number of protons and neutrons within the nucleus. Because the strong force is not all *that* strong, the Pauli principle also works to prevent the existence of a simple nucleus of just two neutrons or just two protons.

If there were no electrical repulsion between protons, there would be no limit to how many nucleons could join together to form a nucleus. There would be a carbon nucleus with six protons and six neutrons (as there in fact is in the real world), a uranium-184 nucleus with 92 protons and 92 neutrons, a nucleus with 300 protons and 300 neutrons, and so on—plenty of room for carolinium and no end of other imaginatively named elements. Visualize a graph in which proton number is plotted vertically and neutron number horizontally. As shown in Figure 1, the nuclei with equal proton and neutron number would lie along a straight line inclined at 45 degrees

---

*We can overlook the tinier quarks and gluons within the neutrons and protons.

and extending on without limit. This is the so-called line of stability. (There would be some stable nuclei with proton and neutron numbers not exactly equal, but we don't need to be concerned with them. They would just change the line of stability into a band of stability.)

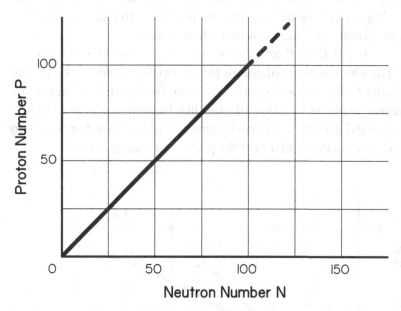

Fig. 1. The nuclear "line of stability" in an idealized world with no electrical repulsion between protons. On this graph—with number of protons P shown vertically and number of neutrons N shown horizontally—stable nuclei show up along a straight line inclined at 45 degrees, reflecting the equal number of protons and neutrons in each nucleus. (The line of stability would actually be a narrow band of stability.)

But there *is* electrical repulsion between protons. What is its effect? For the lightest nuclei, not much. We have deuterium (one proton and one neutron), carbon-12 (six protons and six neutrons), oxygen-16 (eight of each), and neon-20 (ten of each). But as the number of protons grows, the repulsive forces between them begin to work their will. Heavier nuclei are more stable if they have more neutrons than protons. The stable isotope of aluminum, for example, has 13 protons and 14 neutrons. The most common isotope of barium has 56 protons and 82 neutrons, and of uranium 92 protons and 146 neutrons. Beyond a certain point, the electrical repulsion among the protons is more than the strong force can cope with. Element number 83, bismuth, is the heaviest element with any stable isotope at all. Beyond that, there are only unstable nuclei, extending currently up to element number 118 (that is, the element whose nucleus contains 118 protons).

What is the effect of this on our graphical line of stability? The electrical repulsion between protons causes the line of stability to *bend* and to *end*. As shown in Figure 2, as neutrons gain on protons the line of stability becomes a curve, bending toward greater neutron number. And because there are no stable nuclei beyond a certain point, the curve ends.

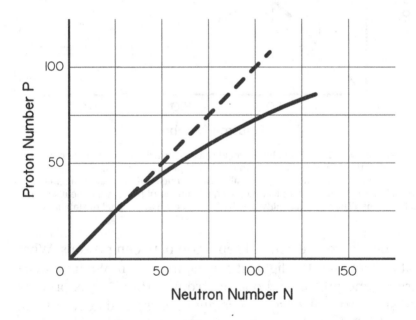

Fig. 2. In the real world, where protons repel one another electrically, the band of stability bends (toward greater neutron number) and it ends (where there are no more stable nuclei).

At this point the reader may reasonably ask: What has all of this to do with fission and fusion? The linkage occurs because as one marches through the elements (or, more exactly, their nuclei) from hydrogen to uranium and beyond, what is called the nuclear binding energy changes in a regular way, driven by the competing effects of the nuclear and electrical forces. The nuclear force wants to pull nucleons together. The electrical force wants to push some of those nucleons—the protons—apart. For the least massive nuclei, containing few protons, the strong force is the clear winner. In the range of intermediate masses, the two forces coexist in uneasy harmony. For the heaviest nuclei, the strong force surrenders to the electrical forces. These nuclei, when they exist at all, live only briefly. Beyond bismuth, all elements are radioactive.

The "binding energy" of a nucleus is the energy needed to pull it apart, to completely disassemble it. Consider the lightest nucleus other than a single proton: the deuteron. It is the nucleus of "heavy hydrogen" and consists of one proton and one neutron. Its binding energy is 2 million electron volts (2 MeV).* The mass of each of its constituent particles is, in energy units, approximately 1,000 MeV, so the binding energy of 2 MeV is about one one-thousandth of the mass of the nucleus. To phrase it differently, the mass of the nucleus is less than the combined mass of a proton and a neutron by about one part in a thousand. Not much, but easily measured.

Let's look next at the nucleus of helium-4, which contains two protons and two neutrons. For this nucleus (which is the same as an alpha particle), the nuclear force easily outcompetes the electrical force, and its binding energy is much greater than that of the deuteron—28 MeV instead of 2. This means that 28 MeV of energy would have to be poured into this nucleus to pull it apart into two protons and two neu-

---

*The electron volt (eV), the standard unit for measuring both mass and energy in the atomic and nuclear world, is the energy acquired by an electron or proton when it moves through a potential difference of one volt.

trons, or 24 MeV to separate it into two deuterons. It also means—here comes fusion—that if two deuterons coalesce to form an alpha particle, 24 MeV of energy will be released. That is exactly what happened in the first thermonuclear explosion, the "Mike" shot in late 1952. (I discuss below why this doesn't happen spontaneously and why it was such a chore to make it happen.)

In considering nuclear energy, it is useful to use as a unit the binding energy per nucleon. This is 1 MeV/nucleon for the deuteron, 7 MeV/nucleon for the alpha particle. Moving toward heavier nuclei, this number grows, but only slowly. It reaches a maximum of about 9 MeV/nucleon for the nucleus of iron, which contains 26 protons (and whose most abundant isotope contains 30 neutrons). By this point, the repulsive electrical force is beginning to overcome the hegemony of the strong interaction. As the line of stability bends with the addition of more nucleons, the binding energy per nucleon declines, sliding back to around 7 MeV/nucleon at uranium.

We call iron the element with the "most stable" nucleus. This means that combining less-massive nuclei to form a nucleus closer in mass to that of iron releases energy. For light elements, *fusion* releases energy. It means, too, that splitting apart a heavy nucleus to create two nuclei closer in mass to that of iron releases energy. For heavy elements, *fission* releases energy.

If fission and fusion release energy when they occur, why don't they occur spontaneously? Why does it take so much scientific and engineering effort to induce these processes to take place?* (Radioactivity, by contrast, *does* occur spontaneously. In fact, nothing can be done to start it or stop it. When

---

*Actually, there is *some* spontaneous fission in nature, but only for rare very heavy elements, and the fission they undergo is really a form of radioactivity. It does not stimulate a chain reaction of the kind that takes place in a reactor and a fission bomb. And once, some two billion years ago, there *was* a naturally occurring reactor on Earth, in what is now Gabon in Africa. But the conditions that set it going are most unlikely to recur.

concentrated, it can be a health hazard, but is otherwise largely harmless.)*

Fusion is inhibited by the electrical repulsion between nuclei (recall that all nuclei are positively charged). One might at first think that when two deuterons find themselves close together, they would fall into each other's arms, combine to form an alpha particle, and release energy. Under normal conditions they can't get close enough for that to happen. "Close enough" really means touching, and that requires that their centers be not much more than 10^{-15} meter apart, a distance that is nearly 100,000 times smaller than the size of an atom. In normal jostling at ordinary temperatures, the mutual electrical push they exert on each other keeps them much farther apart than that. One way to push them close enough to fuse is by using an accelerator to send a beam of high-energy deuterons toward a deuterium target, which is in fact commonplace in the laboratory but releases very little energy by normal standards since the number of particles involved is less than in a minuscule speck of matter. The other way to cause fusion is to heat the material to enormous temperature—tens of millions of degrees. Then thermal energy is large enough to propel some deuterons (or other light nuclei) within reaction range of each other. That is what happens in the center of the Sun and in an H bomb, and what someday may happen in a controlled fusion reactor.

Fission is inhibited in a different way. A fissionable nucleus—that of uranium-235, for instance—is like a boulder in the cone of an extinct volcano. If the boulder can get up and over the lip of the cone, it will tumble down into the valley, releasing energy. If the nucleus can surmount the energy

---

*What we might call an agreeable consequence of radioactivity is that over Earth's history it has created the vast underground reservoirs of helium that we find so useful in technical applications such as cooling superconducting magnets. (An alpha particle brought nearly to rest after being emitted in a radioactive-decay event latches onto a couple of electrons to balance its charge, and, presto, a helium atom is born.)

barrier that holds it together, it will fall apart into two pieces, also releasing energy. The bit of added energy to get the process started, it turns out, can be supplied by a neutron, which can sneak up on a nucleus (since no electric force holds it back), be absorbed by the nucleus, and add some 7 MeV of energy to the nucleus as it (the neutron) is pulled in and joins its fellow nucleons. That 7 MeV of extra energy is enough to allow the nucleus to surmount its energy barrier and come apart. If, in addition to undergoing fission, the nucleus releases more neutrons, these added neutrons can, in turn, stimulate more fission events, and a chain reaction is the result.

Finally, let's look at the energies involved in fission and fusion. When a heavy nucleus undergoes fission, the average binding energy per nucleon goes from 7 MeV (in round numbers) in the "parent" nucleus to about 8 MeV in the "daughter" nuclei, which are closer to the "most stable" iron. The 1 MeV per nucleon of energy that is released multiplied by, say, 236 nucleons in the fissioning nucleus gives a total calculated energy release of 236 MeV (very close to the measured value). Heading to the other end of the periodic table, we find that the fusing of two deuterons to make an alpha particle releases some 6 MeV per nucleon (the 7 MeV per nucleon in the alpha particle minus the 1 MeV per nucleon in the deuterons). For other fusion reactions in light nuclei, such as between protons in the Sun or between deuterons and tritons in an H bomb, the energy release per nucleon is similar. In summary: Although the total energy released *per event* is greater for fission than for fusion, the energy released *per unit mass* is some six times greater for fusion than for fission. This comparison doesn't carry over directly to bombs, since the mass of a bomb includes the mass of ancillary structures, not just the mass of the fuel, but it remains true that an H bomb is more "efficient" than an A bomb. For a given mass, it releases more energy. And both release at least 100,000 times more energy per unit mass than dynamite or TNT.

Radiation as Stuff

L et me explain what was special about the radiation-implosion idea (the 1951 insight of Edward Teller and Stan Ulam that replaced the unattainable runaway Super with the successful equilibrium Super). It has to do with the temperature of matter and radiation and the division of energy between matter and radiation.

That matter has a certain temperature is a familiar idea. The air and the walls in your living room may be at 68 degrees Fahrenheit, or 20 degrees Celsius. The inside of your refrigerator may be at 5 degrees Celsius, your body at 37 degrees Celsius. The temperature of the Sun's surface is 5,500 degrees Celsius. All of these numbers can be rendered, too, in a unit that physicists tend to favor, the kelvin. A temperature in kelvins is the number of Celsius degrees above absolute zero: 293 K for the living room, 278 K for the refrigerator, 310 K for body temperature, and about 5,770 K for the Sun's surface. The center of the Sun is at about 15 million K, and the temperature generated in the core of a fission bomb—which is also roughly the temperature needed to sustain thermonuclear burning—is even greater, some 50 million K.

That radiation may have a temperature is a less familiar idea. The radiation emanating from the Sun's surface mimics the temperature at the surface, or 5,770 K. If it were coming at you from all sides—if you were literally bathed in it—it would vaporize you like a comic-book ray gun. Fortunately, it impinges on you from only a tiny range of angles, so the worst it can do is give you a sunburn.* Filling up all of "empty" space in the universe is radiation, the so-called cosmic background radiation, in which we are indeed bathed. It has a temperature of 2.7 K—very cold indeed but easily measured. These are examples of electromagnetic radiation, which, from a modern

---

*Of course, there is a lot of "best" that sunlight can do, too—warming Earth, energizing plants, and, with the help of solar cells, generating electricity.

perspective, consists of photons running hither and thither. Within your living room, there is actually radiation with the same temperature as that of the walls and the air molecules— tenuous but definitely there.

Not all radiation has to have a temperature. Your mobile phone, for example, is emitting and absorbing radiation, but that radiation has no defined temperature because it is not in *equilibrium* with matter. When matter and radiation are constantly exchanging energy, they can come to a common temperature, just as the walls and the photons bouncing around in your living room do.* As you can surmise, all of this has something to do with thermonuclear weapons. The intention of the classical Super design was for the temperature of the exploding thermonuclear fuel to outstrip the temperature of radiation emitted by the "burning" fuel, so that as much as possible of the energy being generated remained in the matter and as little as possible of it got "lost" in radiation. In the Teller-Ulam design, by contrast, matter and radiation remain in equilibrium and maintain pretty much the same temperature. I discussed at the end of Chapter 1 why this idea didn't surface until 1951, nearly a decade after physicists first started discussing how to make an H bomb.

Now to energy. When something gets hotter, it also gains more energy, but not always in direct proportion—in fact, for radiation, very far from a direct proportion.

Let's suppose that you are a hobbyist who keeps a cubic meter of deuterium (heavy hydrogen) in a large box in your

---

*Compact fluorescent bulbs and light-emitting diodes (LEDs) are often marked with an "effective temperature." If you buy a bulb that is marked "2,700 K," it does not, of course, mean that the bulb reaches such a high temperature. Far from it. It means that the color spectrum emitted by the bulb is similar to that of light from a source at 2,700 K, strong in the red, weak in the blue. If the bulb is marked "5,000 K," its colors are similar to those of a hotter source, strong in the blue, weak in the red. Interestingly, we call the reddish tone of the 2,700-K bulb "warm" and the bluish tone of the 5,000-K bulb "cool."

back yard. Your container is a cube one meter—about three feet—on a side. You are interested in the temperature and the energy of the deuterium and of the radiation that is trapped along with the deuterium gas inside the box. Since you are an amateur physicist as well, you calculate that on a warm summer day, with the temperature around 80 degrees Fahrenheit, or 300 K, and with a pressure in your box of one atmosphere, the total energy in the random motion of all the deuterium molecules in the box is 150,000 joules. (This is the energy of what is called the "translational" motion of the molecules as they bounce around from place to place. It is the energy associated with temperature, and doesn't include the vibrational and rotational energy of the molecules.) What, you then ask yourself, is the energy of the trapped radiation in the box? You can calculate this, too, and you find it to be six *millionths* of one joule—25 billion times less than the energy in the matter. You can also calculate the pressure exerted by that radiation on the walls of its container, and find it to be twenty *trillionths* of one atmosphere. So the radiation in the box (and in your living room) is tenuous indeed.

You could build a fire under your box of deuterium and explore how much the energy of matter and radiation change. The answer would be not much for the range of temperatures you could actually achieve. So, instead, you decide to do a thought experiment—what Einstein and his German-speaking colleagues used to call a *Gedankenexperiment*. What, you ask yourself, would happen to the energy in the matter and the radiation if you raised the temperature of the box from 300 K to 30 million K—not enough to trigger a thermonuclear explosion, but getting close. That would be a factor of temperature increase of 100,000. The energy in the matter, it turns out, would increase by a factor of 400,000 rather than 100,000. This is because at that temperature, where there used to be one deuterium molecule there would now be two deuterium nuclei and two electrons—four particles where there used to be one. The energy in the matter is then some 60 billion joules, or the equivalent of 15 tons of high explosive (0.015 kilotons).

You might then think, "Well, the energy in my hypothetical super-hot box isn't enough to destroy a city, but it could do in a village." But wait. What has happened to the energy in the radiation in this thought experiment? Maybe that changes the picture. It does indeed. The energy in a given volume of radiation goes as the fourth power of the temperature. If you double the temperature (the absolute temperature, in kelvins), the energy in the radiation increases by a factor of sixteen (2 to the 4th power). If you increase the radiation temperature by a factor of ten, the radiant energy increases by a factor of ten thousand (10 to the 4th power). If you increase the temperature by a factor of 100,000, as you have done in your thought experiment, the energy in the radiation increases by the quite enormous factor of 10^{20}, or 100 billion billion. The once-tenuous radiation, which originally accounted for one twenty-five billionth of the energy in your box, now accounts for 99.99 percent of it. The energy in the one cubic meter of radiation at a temperature of 30 million K is, in the units favored by weaponeers, 150 kilotons. And its pressure is correspondingly elevated, to 2 billion atmospheres.

The particular numbers in our thought experiment are not important. What is important is the immensity of radiation's effects—its energy and its pressure—when it is hot enough. At ordinary temperature, radiation is like the pixie dust that was visible only to Tinker Bell and her band of fairies. At the temperatures characteristic of nuclear explosions, radiation is "stuff," full of enormous energy and capable of pushing like a giant piston.

Chapter 7
Going West

I spent most of my boyhood in Kentucky, and, when I was eight and nine, lived for one year in Georgia. In 1942, at sixteen, armed with a "regional scholarship"* from Phillips Exeter Academy, I was off to New Hampshire for the final two years of high school. All of Exeter's students at that time were boys, and almost all of them were from New York and New England. My role, as a southerner, was to leaven the mix. I suppose I did that to some extent. The educational benefit to me was enormous. Before I left Exeter, I knew that I wanted to be a physicist. Going on to Harvard and Princeton seemed more like following a natural course of events than choosing a path.

A slogan that goes back even to well before the 1940s is: Join the Navy and see the world. [1] I did join the Navy—just before turning eighteen and not long before my graduation from Exeter—and I did see at least some parts of the world: Mississippi, Indiana, Illinois, Ohio, and Michigan—although never a ship or a foreign port. The Navy first set about training me to become an Electronic Technician. Recruits who, through testing, showed a scientific or mathematical bent were selected for the program. At the time, radio receivers, transmitters, radar, and sonar were all in a state of rapid development, and all were in need of skilled technicians to keep them running (their vacuum tubes got sick easily). I enjoyed the training. But part way through it, I was offered the option of applying for something called the V-12 program, which meant going to college as preparation for becoming an officer. After managing to finesse

---

*It was a quite remarkable scholarship that covered all expenses, including two round trips per year in a Pullman car.

an eye exam,* I was demoted from Electronic Technician 3rd Class to Apprentice Seaman and sent off to John Carroll University in Cleveland, Ohio, where, among other things, I took a wonderful course in differential equations in a class of three students. From there the Navy sent me to the University of Michigan in Ann Arbor, where, fortuitously, a physics course in which I enrolled required students to be in the lab on the afternoon when Navy drill was scheduled. By this time the war was over, and I waited my turn for discharge. I went off to Harvard in the fall of 1946 being neither an Electronic Technician nor an Officer.

By 1950, when I chose to follow John Wheeler to Los Alamos, I had seen a good deal of the East, the South, and the Midwest, but had never been as far west as the Mississippi River nor to the lands beyond it. This was to be a new adventure.

My transport at the time was a bicycle and a 1931 Packard touring sedan, neither of which seemed right for the wild west. I had purchased the Packard two years earlier for $300 from a fellow graduating senior at Harvard. For that sum I acquired a car of near-limo proportions with a convertible top, a cigarette lighter that extended on a long spooled wire from the dashboard to the remote rear seat, a hood that stretched far out in front of the driver, and two spare tires, one mounted on each front fender. Having two was a good thing, since the tires had a habit of going flat or blowing out at frequent intervals. (On one drive from Boston to New York, I had two blowouts, and my passengers, two students from Wellesley College, got to New York too late to attend the wedding that was their destination. We remained friends.) I had no trouble finding a buyer for my bicycle. To dispose of the Packard, I turned to my mother in Garden City on Long Island (my parents had left Kentucky while I was at Exeter). She was a better business

---

*I listened to the sailors ahead of me in line and inspected the chart with both eyes open. When it came my turn to recite the letters using one eye at a time, I was flawless.

person than I, and she took on the task with enthusiasm. After some word of mouth and some local advertising, she found a buyer for the car. I don't remember how much she got for it, but I do remember that it was more than I had paid for it.

Then came the task of finding a vehicle that *was* suited for the wild west and that fit my budget. I settled on a surplus Army vehicle, a Chevrolet Carryall. It needed only $400 to muster it out of the Army and into my possession. In modern parlance, it was a "crossover" vehicle, a truck-like body on an automobile frame. Its tail gate was supported by chains inside the vehicle, which, when the Carryall was closed up and driven round curves, clanked in a most satisfying rhythm as the chains swung to and fro. It also burned oil, lots of oil, emitting a blur of blue smoke through its exhaust pipe. I turned to a friend from prep-school days, David Nason, who had grown up in Cleveland, Ohio. He had earned a degree in chemistry while I was earning my physics degree, and now worked in a Texaco research lab in Beacon, New York. He was skilled with his hands, and jumped at the chance to overhaul the engine on my new car (or truck)—even though his knowledge of piston rings and crankshafts was, I am pretty sure, only theoretical. Nevertheless, we dived into the task in his garage in Beacon (by this time, I had Princeton's qualifying exam behind me and was largely free of pressure). Dave did the work while I, like an operating-room nurse, handed him the tools he called for, provided rags for wiping things clean, and looked into the engine's innards to make sure nothing extraneous was left behind. Miraculously, it all worked. The reassembled engine did not burn oil, and the vehicle served me well during my year in Los Alamos, including some treks over back-country roads.

Ever constrained by tight finances, I looked for some passengers who wanted to go west and could share expenses. A pair of British graduate students (not physicists) fit the bill perfectly. They wanted to use their summer to visit California, and were happy to see much of the country close up on the way. The front of the Carryall contained a so-called bench seat that could accommodate three people. I had removed the rear seats to make way for a mattress, and on at least one night we slept in the car. Actually, small-town hotels, with typical rates of $2.00 per night for a single room, were within our range. (We avoided motels, whose rooms were priced at $3.00 to $4.00.)

In due course, after making our way along Route 66 through Oklahoma and the Texas panhandle (with no blowouts and no flat tires), we reached Clines Corners, New Mexico, where I had to peel off to the northwest on Route 285 toward Santa Fe and where my passengers could catch a Greyhound Bus to carry them the rest of the way to California.

Some easterners, on first encountering this part of New Mexico, are caught off guard, even made uneasy, by the seeming desolation, the loneliness, the creek beds that contain no water, the palette of every color but green, the persistent sunshine. Where are the trees? they ask. Where are the people? Where are the vibrancy and sounds of the city? These were

New Mexico landscape c. 1950. This view looking east from the end of a Los Alamos mesa shows the road to "the Hill," with Otowi Mesa in the middle distance, and the Rio Grande valley and the Sangre de Cristo mountains in the background.
Photograph by John Pilch, courtesy of Los Alamos Historical Museum Photo Archives.

not my reactions. I fell in love instantly with the unbelievably blue sky punctuated by puffy, hospital-white clouds in the foreground and towering grey thunder clouds in the distance; with the undulating hills and flat-topped mesas; with the crisp, dry air; with the scrub growth and tumbleweeds and road runners. That love affair with New Mexico has lasted to the present day. Even now I go back every once in a while just to "breathe New Mexico's air."

Route 285 brings one into Santa Fe on its southeast edge, into what was once the main wagon trail from the east. One plunges—almost instantly, it seems—from the open skies and beige desert outside the city to Old Pecos Trail and its charming adobe buildings within the city. (This is quite unlike entering the city on its southwest side. That entry point, Cerrillos Road, was, even in 1950, an unattractive line of motels, gas stations, and eateries.) But whatever the entry point, one is led to the Plaza at the center of the city, next to the Pal-

Dorothy McKibbin, the Los Alamos "gatekeeper" for many years, in her Santa Fe office at 109 East Palace Avenue, c.1950. *Courtesy of Palace of the Governors Photo Archives (NMHM/ DCA), negative #030187, and Los Alamos Historical Museum Photo Archives.*

109 East Palace Avenue, Santa Fe, in 1963, the year the office closed. As the sign suggests, the entrance to Dorothy McKibbin's office is to be found inside a patio. *Courtesy of Los Alamos Historical Museum Photo Archives.*

ace of the Governors and close to the St. Francis Cathedral, made famous in Willa Cather's *Death Comes for the Archbishop*. I quickly found Dorothy McKibbin in her inconspicuous office—labeled "U. S. Eng-rs"—at 109 East Palace Avenue, just a block or so from the Plaza. [2] She was the point of first contact for visitors headed for "the hill." As I was checking in with her to get a pass that would let me into the fenced city of Los Alamos, I was conscious of the fact that "Mr. Baker" (Niels Bohr), "Mr. Farmer" (Enrico Fermi), and many other notable scientists had preceded me in that small space on the same mission.

After crossing the Rio Grande and heading toward the higher elevation of Los Alamos, I was stopped, along with other uphill traffic, to make way for a downward-bound convoy of trucks bearing wooden barracks on their oversized trailers—a demonstration that the city's transition from wartime to postwar was still under way. That gave me a chance to chat with some residents, who were evidently happy to live where they did and, as I soon discovered, happy to be sheltered behind a fence.

The road to Los Alamos, looking west, with the curve of the newer road on the right and the sharp switchbacks that trucks had to navigate in the 1940s visible on the left. At the top of the picture in the distance is the Los Alamos airstrip, built in 1947. After it was opened to private pilots in 1968, I often landed there. *Courtesy of Los Alamos Historical Museum Photo Archives.*

Guard gate at Los Alamos town entrance, c. 1950. *Courtesy of Los Alamos Historical Museum Photo Archives.*

Once in the city and installed in a men's dormitory, where I was to share a room with John Toll, I went to see the Wheelers in their "Bathtub-Row" house at 1300 20th Street. This house, and others near it on 20th Street, had been part of the Los Alamos Ranch School before the war and, unlike the housing thrown up by the Army, contained bathtubs.* Next door to the Wheelers, at 1152 20th Street, were the Ulams— Stan, Francoise, and their daughter Claire, then five. (Stan and Francoise Ulam liked to say that their daughter was born in Santa Fe's P. O. Box 1663. In fact a great many babies came into the world in that box.) The Wheeler children at the time were eight, twelve, and thirteen.†

Janette Wheeler and the three children had in fact cut their stay in Paris shorter than planned. The children were pulled from formal education in France and by the end of

---

*The street is now officially named Bathtub Row.

---

†Through the agreeable workings of bureaucracy, I was assigned the bathtub-row house at 1300 20th Street in the summer of 1968, the summer when my seventh child was born. We qualified, apparently, on the basis of family size rather than lab seniority. We enjoyed the large house, the large yard, the Indian ruin in the back, and our neighbors the Ulams.

Rear view of 1300 20th Street, Los Alamos. Behind the three windows on the right is the study where John Wheeler, John Toll, and I did our "Princeton physics" (see page 87). *Courtesy of Los Alamos Historical Museum Photo Archives.*

April 1950 the whole family was settled in Los Alamos. Janette Wheeler did not take to her new home "on the hill." She missed her friends in Princeton and judged the schools in Los Alamos to be inferior to those in Princeton. She saw Los Alamos as a "company town" in which social hierarchies matched professional hierarchies in the lab, and in which newcomers like herself had barriers to overcome. Her attitude played a role in John Wheeler's advocacy later that year for a separate H-bomb group, ancillary to Los Alamos, in Princeton. More on that later.

On my first or second evening in Los Alamos, the Wheeler family took me and John Toll to the eastern end of a mesa just outside the town for a picnic supper and some amateur art. We dutifully tried to sketch the incredible scene before us, with deep canyons near us, Black Mesa visible down in the valley, and the aptly named Sangre de Cristo mountains farther east, indeed blood-red at that time of the day. The children were the better artists, and, in any case, what John Wheeler really wanted to do was to brief me on the state of thermonuclear work at the lab. (John Toll had been there a few weeks and was already embedded in the lab work.) So I got the story in as much depth as a half-hour conversation could cover. I can't remember whether my Q clearance had by that time already been approved, but to Wheeler that must have seemed an unimportant nicety. Within the next few days I was at work in the lab, well briefed and with Q clearance in hand.

As it turned out, there were, at the lab, almost no barriers to open communication among all the scientists, engineers, administrators, and consultants. "Need to know" was very broadly interpreted, leading to an agreeably open—and thus more exciting and more productive—work environment. I remember only one occasion where some information was withheld from me. That was when I was back in Los Alamos for a visit in the fall of 1951, and was denied information on what we knew about additional Soviet nuclear-weapon tests ("Joe 2" and "Joe 3") and how we knew it. By "denied information" I just mean that I was shooed out of an office containing Carson Mark and Hans Bethe and a few others—probably including Edward Teller and John Wheeler—when that topic came up.

John Toll and I were compatible roommates and compatible work-mates. Both of us were used to working long hours, both of us found physics to be exciting, and both of us enjoyed life in the lab and in the town. There was only one small difficulty, which I label "How I learned to hate Martin Agronsky." John Toll, at the stroke of seven each morning, as his alarm clock rang, went from sound sleep to complete alertness in a matter of seconds, seconds in which, while getting out of bed, he reached for the radio and turned it on in order to hear the morning news reported by the gravel-voiced Martin Agronsky. So, as I slowly awakened, I had to endure Martin Agronsky. His program got to be like cod-liver oil.

Toll's interest in world affairs was, at that time, greater than my own. I'm sure, for instance, that he learned about the Korean War on the day it started, June 25, 1950 (while I was en route). It came to my attention only some days later, after I had settled in at Los Alamos.

Chapter 8

A New World

Within days of joining the Los Alamos lab, I was confronted with yet another decision, this one more confounding than the decision whether I should be there at all. All of the lab's three thousand or so employees were presented with a loyalty oath to sign. We were employees of the University of California, and this oath was the same as the version that all UC staff were being required to sign at their respective campuses in California as of July 1, 1950.* [1] The oath had two parts, brief enough to be duplicated in full here.

First, an oath required of all employees of the State of California: [3]

> I do solemnly swear (or affirm) that I will support and defend the Constitution of the United States and the Constitution of California against all enemies, foreign and domestic; that I will bear true faith and allegiance to the Constitution of the United States and the Constitution of California; that I take this obligation freely, without any mental reservation or purpose of evasion; and that I will well and faithfully discharge the duties upon which I am about to enter.

---

*For nearly a year, debate about the oath had been roiling UC campuses. Some faculty resigned in opposition to it, and thirty-one were fired. Among the latter was the UCLA physicist David Saxon, who, after being rehired, became the president of the UC system in 1975. [2] When I was confronted with the oath in Los Alamos, I knew nothing of the controversy it had already generated in California.

Then, a supplement specifically for University employees:* [4]

> Having taken the constitutional oath of office required by the State of California, I hereby formally acknowledge my acceptance of the position and salary named, and also state that I am not a member of the Communist Party or any other organization which advocates the overthrow of the Government by force or violence, and that I have no commitments in conflict with my responsibilities with respect to impartial scholarship and free pursuit of truth. I understand that the foregoing statement is a condition of my employment and a consideration of payment of my salary.

I didn't like the oath and didn't want to sign it, although I could not clearly articulate my reasons for feeling that way. I was, deep down, a libertarian of sorts, and may have carried the "genes" of Quakerism that later became explicit in my life. As it turned out, the oath swept through the Los Alamos lab with barely a ripple. As of the designated deadline for signing, only two lab employees had not signed. One was John Manley, a senior physicist and associate director of the lab. I was the other. Manley had been a key figure in the Manhattan Project, starting with his participation in a meeting in Berkeley in 1942 that got the whole program (including considerations of thermonuclear weapons) started. He left the lab in that summer of 1950 for a department chairmanship at the University of Washington, without having signed the oath. I stood in awe of his principled stand, and had to decide whether to follow his lead (in not signing, not in transitioning to some other good job). John Wheeler did not press me one way or the other. I sat down with Norris Bradbury, the lab director. "Ken," he said (I paraphrase), "I fully sympathize with your position and I agree

---

*In numerous places on the Web, the word "the" is inserted so that the text of the oath reads "... oath of the office."

that the oath is odious. But there is not a thing I can do. My hands are tied. If you don't sign, you will have to be terminated." I wasn't prepared for an abrupt end to my days at the lab. I signed.*

Many years later, in the 1990s, when I was helping John Wheeler prepare his autobiography, I interviewed John Manley's widow, Kay. In the course of the conversation, I mentioned how much I had always admired her husband for his refusal to sign the oath. "Oh, no," she said (or words to that effect), "It's true that John was opposed to the oath, but he already had accepted the job offer from the University of Washington and didn't need to sign. Had he stayed at the lab," (I continue to paraphrase) "I'm sure he would have signed." In fact, John Manley did return to Los Alamos to work there after the oath requirement was dropped, so I shall continue to believe that he acted, at least in part, on principle.

Less than two months after President Truman's statement of January 31, 1950 (or his call for a crash program to build an H bomb, as some interpreted it), the Los Alamos lab went on a six-day week—accompanied by a 20-percent boost in salaries. [5] Teller reports in his *Memoirs* that this change elicited some "grumbling in the ranks." Soon after John Wheeler's arrival in Los Alamos in late February or early March, Teller joined him for breakfast at Fuller Lodge, and, according to Teller, Wheeler "told me that after he had got into bed, he

---

*Years later, in 1971, I joined the University of Massachusetts and was confronted, unexpectedly, with another oath of a similar kind. Most such oaths in other states had been expunged by that time, but not in Massachusetts. This time I said no and didn't bend. When I and two support-staff members at the Boston campus came before a Superior Court judge who was to decide our fate, he decreed that the case should be shelved for consideration at some later time. That later time has yet to come. The case still resides, so far as I know, in the back of some filing cabinet in Boston.

picked up the Gideon Bible on his nightstand; it opened (he assured me with a solemn expression) to the commandment: 'Six days shalt thou labor.'"* [6] That sounds like Wheeler.

By the time I got to Los Alamos in June of that year, whatever grumbling there may have been had subsided—tempered, no doubt, by the salary boost. Many female employees chose to make a mild statement by coming to work in jeans on Saturdays, while continuing to wear dresses and skirts on the other five days. (I am told that when the lab reverted to a five-day week in early 1953, the reaction was muted, the gain in free time balancing the loss of income.)

When I look back now at that 1950-51 year in Los Alamos, I marvel that I had such a full schedule, mixing work and pleasure with no sense of pressure. Such are the blessings of youth. I worked full time at the lab, did a little nuclear research outside the lab, went on weekend outings, and took part in square dancing. I even joined an exhibition dance group in which the men wore black trousers, black shirts, and bolo ties, and the women wore brightly colored flared skirts. We performed our shuffling two-step at events around the state.

The weekend outings—Saturday afternoon to Sunday evening—were mostly with a small group of other young lab employees in my trusty Carryall. We got as far afield as Gallup and Chaco Canyon (nothing is very close in New Mexico). Among my companions on some of these outings was a lovely and very capable "computer" named Miriam Planck (no known relation to the Max Planck who had initiated quantum physics fifty years earlier). During the summer of 1950 she was helping Enrico Fermi with his calculations on the Super (more on those calculations later). We couldn't help noticing that Fermi seemed to need her presence in his office a good deal of the time. Stan Ulam, aware of Miriam's charm as well as her ability,

---

*Lab Director Norris Bradbury officially requested the six-day work week on February 28, no doubt after some weeks of discussion. [7] It was approved by the AEC a week later, [8] and implemented soon after that.

found reasons to drop in on Fermi rather often to check on how the calculations were going. More than forty years later, I had occasion to talk to Miriam on the phone—she was in Los Angeles, I in Philadelphia. She was then Miriam Caldwell, the divorced wife of the physicist David Caldwell, a UC Santa Barbara professor. In our conversation, she told me something about her life since Los Alamos, and added, "Fermi was a very nice man."

In 1950 the lab buildings were still downtown, next to Ashley Pond (a pond named, appropriately, for Ashley Pond, the founder of the Los Alamos Ranch School).[9] They were wooden buildings with names like gamma and gamma prime, situated within security fences. These fences were so close to the buildings that I often imagined that when Fermi was lecturing in his stentorian voice on a summer day with the windows open a spy could easily loiter just outside the fence and collect secret information. The small lecture room doubled as a coffee and social room and had a name, the Reines Raum. How it happened to be named for the physicist (and later Nobelist) Fred Reines and why the Germanic flavor I have not discovered. For a time, John Wheeler and his "boys"—John Toll and I and two additions from the lab, Burt Freeman and Joe Devaney—occupied a single not-very-large office just off the Reines Raum. Wheeler was no prima donna and accepted this in good grace. It actually had some advantages in being next to the room where people gathered mornings and afternoons.

John Wheeler's house was a short walk from the lab buildings. He, John Toll, and I often went there for lunch, and Janette Wheeler made us feel welcome. Wheeler always managed to embarrass me by giving Janette a big warm kiss after lunch as we started back to work—his way of letting her know, I think, that physics wasn't his only passion. Wheeler wanted to turn

Gamma Building in the downtown Los Alamos tech area, probably early 1950s
(Jemez mountains in the background). A chain-link fence provided modest security.
Courtesy of Los Alamos Historical Museum Photo Archives.

one large extra room in the house into a study where he and
John Toll and I could work on our non-lab projects evenings
and the occasional Sunday. All that this required was that three
desks be requisitioned from the Zia Company. Easier imagined
than accomplished. The Zia Company at the time was the sole
landlord and handled all maintenance and the provision of
furniture for short-term rentals. Most residents loved the Zia
Company. Clogged drain? No problem. Call the Zia Company.
Leaking roof? No problem. Call the Zia Company.

So John Wheeler called the Zia Company and asked for
three desks. Sorry, said the Zia person, desks are not on the
approved furniture list. He cajoled. He pleaded. He turned on
his charm. All to no effect. Days went by. Wheeler talked to the
top official of the Zia Company and to administrators at the
lab. Somehow, somewhere, a lock clicked open and three desks
were delivered. Wheeler was, among other things, dogged. So,
for the balance of the year, he and John Toll and I had our
getaway office. I got a little physics done there (I didn't really get going on my dissertation research until 1952). Toll got
more done, because he was already well into his dissertation.

Wheeler no doubt accomplished even more, because he was Wheeler, as dogged in his pursuit of physics as in his pursuit of desks. Going back to World War II, he always tried to keep his "Princeton physics" alive, no matter what the pressures of his government service. [10]

It was during these early months at Los Alamos, especially the summer of 1950, when I got well acquainted with the people who made T Division (T for Theoretical) tick: the regulars such as Carson Mark, Edward Teller, Stan Ulam, Freddie de Hoffmann, and some others who were engaged mainly in fission work, such as Conrad Longmire and Ted Taylor; the consultants and temporary staff who appeared for from a few days to a whole summer; and, of course, my own mentor, John Wheeler. T Division was divided into groups, T-1, T-2, etc. Stan Ulam's group was T-8 and consisted only of himself and the mathematician Cornelius Everett. [11] Most of the H-bomb crew, including Teller, de Hoffmann, Wheeler and his "boys," as well as the distinguished outside consultants, and the brilliant young Dick Garwin (two years my junior) were classified as belonging to T-DO, or T-Division Office, meaning that we were linked directly to Carson Mark. There were few barriers to communication anyway, and even less with this arrangement. There is no place else I could have been, not even back at Princeton, where I could have encountered—not just encountered, but worked closely with—so many brilliant people in such a short span of time.

The "big-three" consultants during that year at Los Alamos were Hans Bethe from Cornell University, Enrico Fermi from the University of Chicago, and John (or Johnny) von Neumann from the Institute for Advanced Study in Princeton. Each of them had made path-breaking advances in science (or, in von Neumann's case, mathematics, although he also contributed to hydrodynamics research and was a leader in computer science before it was called computer science). Bethe, later a Nobelist, was from Germany. Fermi, already a Nobelist,

Carson Mark, 1951. *Courtesy of Los Alamos National Laboratory Archives.*

was from Italy. Von Neumann was from Hungary and would almost surely have been a Nobelist if those prizes covered his fields of accomplishment.

Bethe, oddly, was, to us junior scientists, the least useful of the three, even though he was, by various accounts, [12] including his own, [13] a major contributor to the H-bomb development. On his visits, he would listen carefully as we described what we had been doing. As a general rule, he gave us no feedback. He was also reluctant to give a general-interest talk in the Reines Raum. In our youthful arrogance, we asked ourselves if we were wasting our time talking to him, since nothing seemed to come of it. Presumably he gave his feedback to Carson Mark or to Teller or Wheeler or Bradbury. He surely comprehended in detail everything that was going on.

By contrast, Fermi behaved as an equal colleague with us "youngsters" and always had useful things to say. If we invited him to give a general-interest talk, he was likely to accept and might offer two or three subjects and let us choose. We also enjoyed Fermi's company outside the lab. He loved hiking in the nearby Jemez Mountains and wouldn't admit for a moment that he couldn't hike as fast or climb as briskly as those of us who were half his age. Fermi was a person of habit, who reportedly arose at the same time every day, listened to the same news source every day (not Martin Agronsky, I suspect), and stopped work at the same time every day. If he was subject to high and low moods, it wasn't evident. To me he seemed always upbeat. And Fermi was competitive in every pursuit, be it an uphill climb or a physics problem or a board game. I remember an evening when he and Edward Teller were fiercely playing Parcheesi, each of them acting as if nothing in the world was more important than winning that game. (I was in the game, too, and can't now remember who won.)

Von Neumann was also a delight to be with. His brainpower stuck out in every direction (and his middle had expanded a bit, too). Like Fermi, he soaked up whatever work we described and made useful suggestions about it, also agreeing, more often than not, to give a general-interest talk in the Reines Raum. Von Neumann was an indoor person, preferring an overstuffed chair, some good friends, and an appropriate beverage in his hand to a hike in the mountains. He and Stan

John von Neumann, unknown date. *Photograph by Alan Richards, courtesy of AIP Emilio Segrè Visual Archives.*

Ulam were long-time friends whose acquaintance went back to 1934 when they corresponded about mathematics while Ulam was still in Poland. [14] It was at von Neumann's invitation that Ulam came to America. [15] As it turned out, they had a lot in common. Both made important contributions to pure, abstract mathematics, yet were equally intrigued by quite practical problems. They shared a droll sense of humor and were regular correspondents. [16] In writing to von Neumann, Ulam was not above making unkind comments about Teller. [17]

Fermi and von Neumann were not destined for long lives. Both died at age 53, Fermi in 1954, von Neumann in 1957. Bethe lived—and worked—to age 98. Just a few years before his death in 2005, I heard him deliver an impressive lecture on the theory of supernovas, the subject of his research interest at the time.

In April 1946, four years before I got to know him, von Neumann joined with Klaus Fuchs to offer a design for igniting a Super. In a patent application that they submitted in May of that year, [18] they gave as the title of their invention, "Improvements in methods and means for utilizing nuclear energy." A cover sheet on their patent submission called it, more succinctly and more revealingly, "One proposed design for 'Super.'" It is interesting to visualize their "unlikely collaboration," as Jeremy Bernstein has called it. [19] Klaus Fuchs, lean, exceedingly quiet, always polite, a favorite babysitter among Los Alamos wives during World War II. John von Neumann, well rounded in physique and in interests, an ebullient bon vivant, probably very far down anyone's list of prospective babysitters. What they had in common was intellectual brilliance—and curiosity.

Their patent application says that their invention was conceived on April 18, 1946 and that it was disclosed at that time to Edward Teller and Robert Serber. As it happens, Thursday, April 18, was the first day of a three-day conference at Los Alamos, chaired by Teller and devoted to prospects for the Super.[20] Earlier, just after the end of World War II, Enrico Fermi had delivered a series of lectures at Los Alamos on the Super as then conceived. (Thanks to Fuchs, notes on Fermi's lectures promptly reached the USSR, and were eventually published in Russia in the original English as well as in translation.[21]) Now, in April, thirty-one people gathered to pull together what was known about thermonuclear burning and to look further ahead. Among the participants, in addition to Teller, Serber, von Neumann, and Fuchs, was Stan Ulam.[22]

On the patent application, von Neumann's name is first. Four years later, Fuchs, while being interrogated in England after his arrest for spying, claimed that the idea was his.[23] We will never know. I am free to imagine the following conversation during a coffee break on April 18.

> *Von Neumann:* Klaus, I'd like to learn more about your idea for igniting DT using radiation. Let's get together for a drink after today's session.

> *Fuchs: Ja, bestimmt,* Johnny. I'll see you at the Lodge.

Von Neumann may have known more about the paper work involved in a patent application.

As the imaginary conversation above suggests, their idea was to use radiation from a fission bomb to ignite thermonuclear fuel, but through a different process than the one later envisioned by Teller and Ulam. I save the details of the Fuchs-von Neumann invention for the next chapter.

Chapter 9
The Classical Super

According to Tom Glazer and Dottie Evans in their album "Space Songs," [1]

The Sun is a mass of incandescent gas,
A gigantic nuclear furnace.

It is that, and it is an H bomb. In the Sun, just as in an H bomb on Earth, hydrogen nuclei fuse to make helium nuclei, releasing energy. One big difference between the solar H bomb and the terrestrial H bomb is speed.* The process in the Sun is slow, *very* slow. It will take about ten billion years for the Sun to consume most of its hydrogen fuel (it's about half gone now). An H bomb on Earth runs through a sizeable part of its fuel (likely to be a compound of lithium and deuterium)—in little more than a millionth of a second. What accounts for that difference in time scale is, in a word, gravity. All that holds the terrestrial H bomb together, if only for a small fraction of a second, is inertia. With stupendous acceleration, the bomb's case gains speed and blows apart. By contrast, the Sun is held together, almost "forever," by gravity. The thermonuclear explosion taking place in the center of the Sun, although sufficient to transform more than four million tons of mass into energy every second and sufficient to bathe the Earth in life-sustaining light, is insufficient to overcome the gravity that holds it all together. So the Sun keeps on "exploding," and we reap the benefit.†

---

*Another difference is that the particular nuclear reactions are different, but both in the Sun and on Earth the net effect is the same: the transformation of hydrogen into helium.

When, in the summer of 1942, Robert Oppenheimer convened a group of nine theoretical physicists (seven already recognized for their achievements,[‡] and two bright "youngsters"[**]) to spend a few weeks together in Berkeley reflecting on how nuclear physics might be applied to war, their subject matter was relatively new in the firmament. Nuclear fission dated only from December 1938, and was a big surprise when it made its appearance. Nuclear fusion had been discussed earlier but came of age as an explanation of stellar energy and as quantitative science only with the work of Hans Bethe in the late 1930s. So here was this small band of "luminaries,"[2] as Oppenheimer called them, charged with the awesome task of considering how to apply the new knowledge of nuclear physics to the pursuit of war.

Fission had no sooner been discovered than physicists recognized its potential to generate large quantities of energy on Earth—either explosively or under controlled conditions. Such energy generation requires a "chain reaction." Enrico Fermi and Leo Szilard, working at Columbia University in New York, had demonstrated the likelihood of such a chain reaction soon after Niels Bohr brought the news of fission to America in January 1939. On average, they found, a fission event stimulated by a single neutron generates more than one additional neutron, providing the possibility of an ever-increasing cascade of energy release. Later that year Hitler launched World War II. In December 1941, the United States entered the war.

[†]The Sun's energy output is equivalent to about ten billion terrestrial H bombs per second. Yes, we do call that slow.

[‡]The six notables besides Oppenheimer were Felix Bloch, Hans Bethe, Emil Konopinski (later my colleague at Indiana University), Robert Serber, Edward Teller, and John Van Vleck. Also on tap for some of the discussions were the local Berkeley experimental physicists John Manley, Edwin McMillan, and Emilio Segrè. (The group included no less than five future Nobelists: Bloch, Bethe, McMillan, Segrè, and Van Vleck.)

[**]Stanley Frankel and Eldred Nelson, both in their early 20s.

Add to these events the deep concern about German prowess in science, and it is easy to see why Oppenheimer's study group was convened. And easy to see why nuclear fission was the group's initial focus—a focus that, before long, led to "Project Y" in Los Alamos and to a fission bomb, or "A bomb."

But the Berkeley gathering was barely under way when the subject of fusion intruded.

Within just a few days, the participants reached the conclusion that a fission bomb could probably be built—a conclusion based on what they already knew or could calculate about the mechanism of fission, the diffusion of neutrons through matter, the likelihood of random neutrons in the environment, and, not incidentally, how fast it might be possible to fire one piece of uranium or plutonium against another. Serber, Frankel, and Nelson had already been working on these questions in the weeks preceding the conference. Now they went into another room, so to speak, to address in more detail the question how a fission bomb might actually be constructed. Serber thought so carefully about that question that by the time the Los Alamos project got under way in March 1943, he was the

Robert Serber, unknown date.
Courtesy of AIP Emilio Segrè Visual Archives.

leading expert on the prospective A bomb and was tapped to give introductory orientation lectures to newly arrived scientists and engineers on "the hill"—lectures that are still famous and still in print under the title *Los Alamos Primer*.[3] So oracular was Serber that when notes on his postwar lectures on nuclear physics at Berkeley were later gathered together, they were titled *Serber Says*; an updated version is still in print.[4]

Enrico Fermi was always turning ideas over in his head. One day in the fall of 1941, walking with Edward Teller and others in New York toward Columbia University's Pupin Hall (the physics building where the group was working on nuclear fission), Fermi asked Teller if he thought an atomic explosion might be used to produce a thermonuclear reaction—that is, an H bomb.[5] Fermi suggested that if deuterium rather than ordinary hydrogen were used, the chances of making that happen might be improved. Teller tells us in his *Memoirs* that a week or two later, while the two of them were taking a Sunday-afternoon stroll, he explained to Fermi why it wouldn't work.[6] Teller had hit on the idea that was in fact destined to dog the classical Super for the next ten years. Too much en-

Emil Konopinski, unknown date.
Courtesy of AIP Emilio Segrè Visual Archives, Physics Today Collection.

ergy, he reasoned, would be radiated away, escaping the mass of deuterium instead of heating it to a temperature at which thermonuclear burning could be sustained.

The problem bothered Teller, though. A few months later, in the spring of 1942 (not long before Oppenheimer's Berkeley gathering), Teller, newly arrived in Chicago, sat down with another new arrival there, Emil Konopinski, to see if the two of them could prove mathematically that the process envisioned by Fermi was indeed impossible, as Teller at first supposed, or if there was some flaw in his argument such that a thermonuclear reaction, initiated by a fission bomb, might in fact be possible. [7] They concluded that they could *not* prove the impossibility of the H bomb, that perhaps with enough tinkering of design, it might be possible. Teller became a convert and never let go of the idea.*

Fusion probably more than fission was on Teller's mind in early July 1942 when he and his wife Mici along with Hans and Rose Bethe boarded the train in Chicago for California. Teller had the opportunity on the train ride to explain his thinking to Bethe and to convince Bethe that the subject of an H bomb deserved more study. As indeed it got. After Oppenheimer's group reached an upbeat assessment of the prospects for the fission bomb just a few days into the Berkeley meeting, the focus of attention shifted to H bombs, whose prospects were a good deal fuzzier, but which offered wonderful challenges for the physicists' minds. [8] The word "Super" was coined at that meeting, [8] and remained current although later modified to "classical Super" after the radiation implosion idea was born.†

On December 2, 1942, less than half a year after the Berkeley conference and just short of a year after the attack on Pearl

---

*Teller's love affair with the H bomb extended into his old age, when he advocated its use for peaceful purposes such as digging canals.

---

†The word "Super" was intended to be descriptive. When the term "gadget" was later coined for the fission bomb, or A bomb, it had just the opposite intent, *not* to be descriptive.

Harbor, Fermi's "pile"—the first nuclear reactor—went critical in Chicago. At that time, plans were already well advanced for far larger reactors in the state of Washington to produce plutonium and for an isotope-separation plant in Tennessee to produce uranium-235. By then, the Manhattan Project had been created with General Leslie Groves at its head,* [9] and the Ranch School in Los Alamos, New Mexico had been selected as the place to design and build a nuclear weapon. [11]

When Project Y got going in Los Alamos in March 1943, Teller and perhaps a few others wanted the lab to pursue both the fission and fusion tracks in a balanced way. As it turned out, fission got a lot of attention and fusion very little during the war years and even afterwards, up to the first Soviet A-bomb explosion in late summer 1949. To use a slang term, thermonuclear weapons were put on the back burner, and they stayed there for more than seven years following the 1942 Berkeley conference. The Los Alamos Lab's director, Robert Oppenheimer, and the head of its Theory Division, Hans Bethe (holding the T-Division post that Teller had hoped for [12]), realized that the fission bomb posed challenge enough, and that, in any case, no fusion bomb would work without a fission-bomb trigger. Teller was, however, allowed—perhaps even encouraged—to pursue his H-bomb ideas on his own. In his "Many People" article, [13] Teller puts it this way: "In spite of the urgency of the situation, Oppenheimer did not lose sight of the more distant possibilities. He continued to urge me with detailed and helpful advice to keep exploring what lay beyond the immediate aims of the laboratory." Teller adds that Oppenheimer's advice was not easy to take, since it would have been easier for him (Teller) to participate in the lab's central mission. This may be disingenuous, but it's also quite possible

---

*The Manhattan Engineer District, as it was officially called, was established in August 1942 and briefly headed by Col. James Marshall. Groves was named its director in September 1942 and, at the same time, promoted to the rank of Brigadier General. [10]

that Teller believed what he was writing when he wrote it. He did make contributions to the fission bomb, but it seems clear that his heart belonged to fusion. He worked at first in T Division, then was transferred to F Division, headed by, and named for, Fermi. [14] F Division was to "consider issues outside the main project." [15]

In any event, through whatever combination of the lab's encouragement and its acquiescence, Teller did keep a theoretical H-bomb program alive during World War II. He undoubtedly got Fermi involved in some way. That involvement is clearly shown in Fermi's 1945 lectures on the Super, to which I alluded near the end of the previous chapter. [16] These lectures provide a beautiful view into Fermi's characteristic clarity and thoroughness, mixing basic theory, practical application, and numerical examples.

Fermi ruled out an equilibrium Super, as did everyone else for the next five years. Here is what he said: "If thermal equilibrium between particles and radiation were established, it would be impossible to heat deuterium to required temperature. In actual fact there will be no thermal equilibrium and we have to consider rate of energy transfer from electrons to radiation."* In other words, the runaway Super is the only option. Fermi went on to calculate ignition temperatures for deuterium with various admixtures of tritium. He noted that a 50-50 DT mixture would burn if exposed to the temperature of an exploding fission bomb, but that with a much smaller, and more practical, admixture of tritium, the ignition temperature would be out of the range of what could be directly supplied by a fission bomb. Fermi even made a crude drawing of a possible two-stage fission-fusion bomb. These lectures set the whole tone of thermonuclear work until the Teller-Ulam concept of the highly compressed equilibrium Super arrived on the scene in 1951.

---

*The omission, twice, of the word "the" makes one wonder if the final transcriber was Russian.[16]

The von Neumann-Fuchs invention, proposed not long after Fermi's lectures (and which I introduced at the end of the previous chapter), involved the use of radiation from a fission bomb to help ignite thermonuclear burning, but the mechanism was indirect, not direct.

They specified a four-stage device, consisting of

- a "detonator" (a fission-bomb trigger)
- a "primer" (a 50-50 DT mixture, the most easily ignited hydrogen fuel)
- a "booster" (D with a 4-percent admixture of T to facilitate its burning)
- a "main charge" (pure D, with no limit on size, that might burn if suitably ignited with the primer and booster)

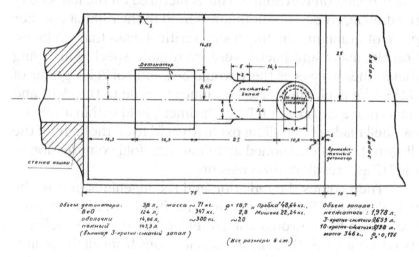

This drawing from a 1948 Soviet publication [16] is probably lifted from the 1946 von Neumann-Fuchs patent application, with Russian legends added. The drawing and other details are presumed to have been supplied to the USSR by Klaus Fuchs. A gun-type fission bomb is on the left, a capsule of thermonuclear fuel (DT) on the right.

The drawing I present here is from a Russian publication dated April 16, 1948. [17] The labels are in Russian, but the text is in English as well as Russian. It is widely assumed that this document and its drawing are taken from what Fuchs supplied to the Soviets in March 1948, but there is no way to be certain of this. [18] (There is still no English version published in the United States.) If Fuchs transmitted information about his joint invention with von Neumann to the Soviets in 1946, as he could have, it may have been more offhand, less detailed than in 1948. In any case, it seems not to have had a measurable effect on the Soviet program. Nor, for that matter, on the American program. I don't recall it being discussed—although, as I shall explain in the next chapter, the Greenhouse George shot in May 1951 did have some features in common with what von Neumann and Fuchs had proposed.

For the "detonator" (the fission-bomb trigger) von Neumann and Fuchs proposed a gun-type weapon similar to the one dropped on Hiroshima. This is pictured on the left side of the drawing. A piece of uranium is fired to the right at another piece of uranium, creating a supercritical mass that explodes. Von Neumann and Fuchs were even quite specific in saying that in the two pieces there would be a total of 71 kilograms of U235. The exploding mass continues to slide to the right and slams into a capsule of DT (the "primer") held within a spherical shell made of beryllium oxide (the circle on the right of the diagram). They envisioned an initial threefold compression of the DT as a result of this collision.

Then comes the radiation and the ingenious part of the von Neumann-Fuchs invention. The flood of radiation from the fission bomb would completely ionize the BeO shell and its DT contents—that is, strip all electrons from all the atoms, leaving a plasma consisting of electrons and atomic nuclei. Since beryllium is the fourth element in the periodic table and oxygen is the eighth, where there had been one Be atom and one O atom there would now be fourteen particles (four electrons and one nucleus from the Be atom, eight electrons and one nucleus from the O atom). Inside the capsule, however, one D atom and one T atom would together yield only four particles (two each). So the BeO plasma, now having many more particles per unit volume than the DT plasma, would experience a much greater increase in its pressure and would exert a mighty compressive force on the DT. Von Neumann and Fuchs estimated that this one-two punch of a direct hit by the fission bomb followed by a radiation-induced implosion of the BeO capsule would result in a ten-fold compression of the "primer"—enough, they concluded, to get thermonuclear ignition.

The burning "primer," they then concluded, could ignite the "booster," which, in turn, would set the "main charge" aflame, with multimegaton consequences. Again being quite specific, they suggested that the "main charge" would reside in a cylinder about two feet in diameter.

There are two important differences between the von Neumann-Fuchs invention and the Teller-Ulam scheme that eventually proved successful. First, in the von Neumann-Fuchs scheme, radiation is used to heat material, not directly to compress it. You could say that von Neumann and Fuchs were considering the energy *content* of the radiation, while Teller and Ulam were considering its *pressure* (pressure augmented, even in their scheme, by a hot plasma and evaporation, or ablation, from the imploding container). Second, von Neumann and Fuchs assumed that the "main charge"—the large cylinder of deuterium—would remain essentially uncompressed. Their device would still be a classical Super, just set burning in a new way (a way that, in fact, would not have worked).

Teller, like most other Los Alamos physicists, left the "Hill" in 1945 following the end of World War II. But in 1949, he came back—before the Soviet nuclear explosion in August 1949, before the GAC meeting that October, before the Truman statement of January 31, 1950, before the six-day work week—for what was to be a one-year leave of absence from the University of Chicago, with thermonuclear weapons on his mind. Once the Soviet bomb was detonated and detected, once Truman issued his statement, the stars and planets realigned. Suddenly Teller was not an outlier at the lab. He was in the middle of the lab's priorities. By early 1950 John Wheeler had joined him, soon to be followed by Lothar Nordheim, [19] John Toll, and me, plus a pair of reassigned young physicists in the lab. Wheeler stayed a year, then headed the ancillary Project Matterhorn at Princeton University for two more. Nordheim stayed for two years, then went back to his academic post at Duke University. In 1951 Teller, piqued at being passed over as head of Los Alamos's thermonuclear program, returned for a time to the University of Chicago. But he devoted almost the entirety of his career after 1949 to thermonuclear weapons—some of that career back at Los Alamos but most of it at Livermore,* a new weapons lab for which he had successfully campaigned.

By the time that H-bomb work shifted into high gear in early 1950, scientists had a pretty clear idea of what the Super might contain and what it might look like, just no idea whether it would work. Within a heavy container, perhaps made of steel and either spherical or cylindrical, would be the thermonuclear fuel, some part of which would, in one way or another, be brought to a temperature of tens of millions of degrees by the explosion of an adjacent fission bomb. The fuel could be liquid—and *very* cold—deuterium (heavy hydrogen), or deuterium salted with expensive and hard-to-produce tritium (still heavier hydrogen), or a solid substance, lithium hydride, as much as possible of it composed of the isotopes lithium-6 and deuterium (thus often called lithium-6 deuteride). Whatever the fuel, if the device was to work, the temperature would need to be maintained or increased after it was ignited, just as

---

*Livermore, founded in 1952 as a branch of Berkeley's Radiation Laboratory, became the separate Lawrence Livermore National Laboratory in 1980.

in a fireplace or coal-burning stove. If too much energy was lost, by radiation or in some other way, the thermonuclear flame would not propagate. The device would fizzle.

Just such a fizzle was feared. As I discussed back in Chapter 1, the theorists were concerned that with little or no compression too much energy might be radiated away, leaving behind not enough to keep the flame going.

This was the situation in early 1950. Since computers were then mostly women with desk calculators, and scientists had only their slide rules, it had not yet been possible to carry out calculations with enough precision to reach a definite conclusion about whether any version of the classical Super would work. Some people—including the distinguished mathematician John von Neumann, his wife Klari, the Los Alamos physicists Harris Mayer and Nick Metropolis, and Cerda and Foster Evans (a husband-and-wife team from Los Alamos)—made a herculean effort over the years 1946 to 1950 to wring some answers from what was the world's first true computer, the ENIAC—built in Philadelphia and then stationed in Aberdeen, Maryland. [20] But that computer, huge of girth and small of brain, was not up to the task, and prospects for the Super remained cloudy. Refined calculations, both with people and with better machines, were evidently necessary.

The people included no less than the esteemed mathematician Stan Ulam and the visiting physicists John Wheeler and Enrico Fermi. The machines included an upgraded ENIAC in Maryland and IBM CPCs (card-programmed calculators) in various locations, with MANIACs in Princeton and in Los Alamos under construction. The fuel of choice for calculations was deuterium with various admixtures of tritium. Lithium-6 deuteride, although more promising in the long term, was harder to analyze. The calculations were challenging enough just with D and DT fuels.

When I arrived in Los Alamos in late June 1950, Stan Ulam and his associate Cornelius Everett were wrapping up a series of calculations using no more machinery than a desk-top calculator. In the calculations, they subjected a quantity of deuterium, with varying enrichments of tritium, to an initial high temperature of the kind that might be provided by a fission bomb and traced what happened. Did the thermonuclear flame spread or die out? Did the temperature increase, hold its own, or diminish? They found that the temperature did not increase and the flame did not propagate, even with unrealistically large admixtures of the tritium elixir. [21]

Whether adding tritium is "realistic" or "unrealistic" is dictated by its cost. Tritium, with a twelve-year half life, is not found in nature in other than trace amounts. Making it takes a large and expensive reactor, one that, if not making tritium, could be making plutonium. So fueling a Super with, say, a 50-50 mixture of D and T appeared to be out of the question. Nevertheless, reactors did eventually make substantial quantities of tritium, which found use in both fission and fusion bombs—and, more to the point, combined fission-fusion bombs. I remember sitting in on a 1950 meeting in Los Alamos with a large map of the United States spread out before us. The fingers of the decision makers in the room pointed toward Aiken, South Carolina. Action was swift thereafter. In October 1950, President Truman asked DuPont to build reactors there, as it had done in Hanford, Washington, during the war. [22] Construction started in 1951, and eventually five reactors at what was called the Savannah River Site churned out both tritium and plutonium. [23] (Incidental to that task, the reactors also created neutrinos—actually antineutrinos—by the countless billion, and it was at Savannah River that Fred Reines and Clyde Cowan in 1956 achieved the first detection of these stunningly elusive particles. Reines—once my next-door neighbor in California—went to Stockholm in 1995 to be recognized with a Nobel Prize.)

At around the time that Ulam and Everett were wrapping up their calculations (and annoying Teller, whose body English had not been sufficient to stimulate encouraging results), Enrico Fermi, in Los Alamos for the summer, began a related series of "burning" calculations with the assistance of the talented Miriam Planck, and with Ulam playing an advisory role. Planck, using equations provided by Fermi, laboriously filled in the cells in a chart (literally a spreadsheet on paper), following the thermonuclear reactions in time and in at least one dimension (either radially in a sphere or axially in a cylinder). Their results were no more encouraging than those of Ulam and Everett.

I familiarized myself with what Ulam and Fermi were up to, and at the same time mastered the behemoth CPC with its instructions encoded on a stack of punched cards. There were several CPCs in Los Alamos and more at Sandia Labs in Albuquerque as well as in New York. For about twelve to eighteen months in 1950-51, the CPCs played an important role in thermonuclear calculations, serving as a bridge between the desk calculators and slide rules of the past and the electronic computers of the future. (More on those computers later.)

To some at the lab, these discouraging results meant that the Super as then envisioned was unlikely ever to become reality. To others, such as the driven optimists Teller and Wheeler and their retinue of young researchers, including me, the results only told us that more and better ideas were needed—on the geometry, the choice of fuel, the mode of ignition, the deployment of fission and fusion components. As I described in Chapter 1, one big, crucial idea—radiation implosion—did appear on the scene early the next year (March 1951) and changed everything. (Teller did say later that the Ulam-Everett calculations "proved" that the classical Super would not work, [24] although he reached no such conclusion at the time.)

Chapter 10
Calculating and Testing

I n the nine months or so between the Ulam-Everett-Fermi-Planck* gloom and the Teller-Ulam rosy dawn, how did we occupy ourselves?

In two ways. First, by hammering away with new and more elaborate calculations on various versions of the Super and on alternative thermonuclear devices such as the layered alarm clock. Teller and Wheeler had been asked to gather together the results of these new calculations, along with summaries of all that had come before, for presentation to the September 1950 meeting of the AEC's General Advisory Committee in Washington—an "as of this time" overview of all that was known or projected about thermonuclear weapons.

The request for such a report came in the form of a letter of July 19, 1950 from Oppenheimer, the GAC Chairman, to Norris Bradbury, the Lab Director. [1] In the letter, cleared in advance with AEC Chairman Gordon Dean, [2] Oppenheimer wrote:

> They [four GAC members who visited Los Alamos earlier in July] have been deeply interested in the new, more quantitative work bearing on the superbomb, and profoundly impressed with the progress that the Laboratory has made in coming to grips with these problems. We all feel that it would be of the greatest help to the whole effort if we could have, in time for our next scheduled meeting on September 11, a quite informal account of the findings and views of the Laboratory on the superbomb problem.

---

*And aided by another "computer," Josephine Elliott.

After noting that the Lab might consider September too early for such a report because "new calculations will be in progress, many questions will be quite unanswered," he added:

> Nevertheless, we believe that the progress and clarification already achieved are so important to sound overall planning and evaluation that we feel it proper to press you to accede.

Oppenheimer knew quite well the state of the "super-bomb" work—that the outlook was very discouraging, that the prospect for what we later called the classical Super was dismal, all the more so after the most recent calculations. Was this letter therefore entirely disingenuous, intended to engineer embarrassment for Teller? I don't dismiss this as a possibility. These two titans had been at odds at least since 1943, when Teller wanted the wartime Los Alamos Lab to give more attention to thermonuclear weapons. And as recently as September 1949, when, in reaction to the just-announced Soviet atomic bomb, Teller phoned Oppenheimer to ask "What do we do now?" and Oppenheimer answered "Keep your shirt on." [3]

I leave it to historians of science to unravel Oppenheimer's motivations and decipher this letter (if indeed it needs deciphering). Neither I nor any other junior member of the team saw any evidence that it aroused suspicion in Teller or Wheeler. If it did, they kept it to themselves. As far as we could tell, they welcomed the request and the opportunity it afforded to summarize all that they knew or surmised about thermonuclear weapons. The rest of us—which means principally John Toll, Burt Freeman, and I—jumped in and found it exhilarating to be part of the effort to organize an overview of past results and possible future directions.

We had only a few weeks to throw the report together. By the time it was completed, in August 1950, [4] it had a slight obesity problem and we called it the "telephone book." Even though it presented disappointing past calculations, the report, overall, was anything but pessimistic. Teller and Wheeler, by nature optimistic, still had high hopes and wanted to encourage the GAC to recommend continued high-priority work on thermonuclear weapons. After reviewing the report, the GAC said: [5]

> ... we are happy to note that great progress has been made at Los Alamos in setting up quantitative methods for coping with the difficult problems of the hydrogen bomb; we urge the Commission to support these efforts by making computational machinery available to the laboratory.

Our second activity in that interim period was calculating expected results of newly scheduled tests of thermonuclear principles in the Greenhouse test series scheduled for Enewetak* in May 1951. I will discuss these in the latter part of this chapter.

Regarding Greenhouse, the GAC, in the report of its September 1950 meeting, had this to say: [6]

> We wish to make it clear ... that the test, whether successful or not, is neither a proof firing of a possible thermonuclear weapon nor a test of the feasibility of such a weapon. This test is not addressed to resolving the paramount uncertainties which are decisive in evaluating the feasibility of the Super.

Thank heaven all atomic nuclei are positively charged. That means, as discussed in Chapter 6, that they all repel one another and keep their distance—at least at temperatures less than some tens of millions of degrees.[†] If any nucleus were

---

*Spelled Eniwetok at that time.

[†]Kelvin degrees, not meaningfully different from Celsius degrees when you are talking millions. (See page 67.)

neutral or negatively charged, it could sneak up on one of its positively charged brethren and react with it or unite with it, producing a mini-explosion and releasing a great deal of energy—at least by atomic standards. If the nuclei belonged to relatively light elements, below iron in the periodic table, the result of their union would be an energy release from a hundred thousand to a million times more than is typical for a pair of atoms in a chemical reaction. If the universe were made of nuclei that could so easily get together, it would contain no cold cinders like Earth or Mars. There would be only glowing suns and glowing planets.

Higher temperature means higher speed for the atoms and nuclei. At high enough temperature, such as exists at the center of the Sun or in an exploding H bomb, some of the caroming nuclei, as I discussed in Chapter 6, can surmount the barrier of electric repulsion between them and, so to speak, reach out and touch one another. Think of yourself accelerating a car on a level roadway, then shifting into neutral and seeing how far up a slope you can roll. If the slope is Pike's Peak, it doesn't matter whether you approach at 10 miles per hour, or 50, or 200. You still won't reach the top. That is like nuclei at ordinary, or even somewhat elevated temperature. Now imagine that you could increase the car's energy a thousand-fold. Then you might have a shot at rolling to the top of the mountain. That is like nuclei at millions of degrees.

What happens then when the nuclei do "touch"? Consider a pair of deuterons. There are two possibilities and they occur with nearly equal probability. Written like chemical formulas, those reactions are:

$$D + D \rightarrow {}^{3}He + n \quad \text{(energy release 3.3 MeV)}$$

$$D + D \rightarrow T + p \quad \text{(energy release 4.0 MeV)}$$

The two reacting deuterons contain a total of two protons and two neutrons. It is easy to see that the products of their reaction, either a helium-3 nucleus and a neutron, or a hydrogen-3 nucleus (denoted by T for triton) and a proton, also contain two protons and two neutrons.

Those energy releases amount to about 1 MeV per nucleon (n or p), about the same as the energy release per nucleon in the fission of a uranium nucleus. But there is more. The triton created in half of the D-D reactions reacts with a deuteron to create an alpha particle, or helium-4 nucleus. The reaction is:

$$D + T \rightarrow {}^4He + n \quad \text{(energy release 17.6 MeV)}$$

The neutron that emerges from this reaction takes 80 percent of the energy that is released, about 14 MeV. Since 14 MeV is considerably greater than the energy of any other particle that is involved, this particular neutron is relatively easy to detect and provides a "signature" of fusion (or, if you prefer, a smoking gun). It is this DT reaction that inspired the graphic used in this book's section dividers.

With this added reaction, the energy release per unit mass is more than twice that of fission. And the cost of the input material, deuterium, is much less than the cost of uranium-235 or plutonium-239. On top of that, there is no limit to how much deuterium can be stored in one place. Just as for wood or coal or oil, there is no critical mass. And the neutron furnished by the DT reaction has enough energy (14 MeV) to stimulate fission in uranium-238, not just uranium-235. So it is easy to see why, right from that 1942 conference in Berkeley, thermonuclear weapons were seductive for the physicists involved in weapons design.

It turns out that the DT reaction not only releases much more energy than the DD reaction, it also proceeds more vigorously—that is, with higher probability and at somewhat lower (although still enormous) temperature. That is why Ulam and

Fermi investigated the possibility of starting with some tritium in the fuel, not waiting for it to be formed in the DD reaction. It would not be surprising if they also considered a scheme like the one first imagined by von Neumann and Fuchs—a Super containing some tritium in its ignition region and only deuterium in its propagation region (the thermonuclear equivalent of a fire started with kindling and kept going with large logs).

Theorists like deuterium because it is a fairly simple substance and this facilitates computation. Engineers and Air Force Generals don't like deuterium. It is a gas at ordinary temperature, and can be made dense only by liquefying it at a temperature of 23 kelvins above absolute zero, or 250 Celsius degrees below zero. The theorists' point of view prevailed for the first large-scale thermonuclear test (Mike, November 1, 1952). The Mike device weighed 62 tons [7] and was a very long way from being portable on an airplane or missile (or even a truck). From soon after Mike, and up to the present day, so far as I know, the thermonuclear fuel of choice has been lithium-6 deuteride, or $^6Li^2H$. This substance has two advantages over liquid deuterium. First, it is a solid at ordinary temperature. You might say that the lithium has provided a matrix that enables the deuterium atoms to cozy up to one another. Second, the lithium adds another useful nuclear reaction to the three I described above. It is

$$_3^6Li + n \rightarrow T + {_2^4}He \quad \text{(energy release 4.9 MeV)}$$

The subscripts show the atomic number (the number of protons in the nucleus, which is the element's place in the periodic table), the superscripts show the total number of protons and neutrons in the nucleus. As before, T stands for triton, the nucleus of tritium. The neutron on the left side of the reaction equation results from the DD reaction (the first branch in my list above). Only computational complexity kept us from considering lithium-6 deuteride for the first thermonuclear tests.

Nowadays, to the extent that there are any new nuclear weapons in the United States arsenal,* they must be designed, built, and deployed without testing (a task not quite as formidable as it might at first sound thanks to the vast store of knowledge from past testing and the power of modern computers).

In the early days of nuclear weapons, on the other hand—during and soon after World War II—testing was considered essential, starting with the Trinity test in Alamogordo in July 1945. (To be sure, the Hiroshima bomb was used without testing, a choice dictated by its conservative design and, significantly, by the short supply of uranium-235 at that time.) Immediately after the end of the war, scientists and managers at Los Alamos began planning for tests as new designs took shape.

The first of the post-war tests took place at "Crossroads," in July 1946, less than a year after the war's end. There, at Bikini atoll in the Marshall Islands, two bombs were exploded, one of them under water. These explosions were followed by three more at "Sandstone" at the Enewetak atoll, also in the Marshall Islands, in April and May 1948. [9] Planners in the United States, with much future testing in mind, were all too aware of the costs of conducting tests many thousands of miles away (more than 10,000 people, most of them military personnel, took part in Sandstone [10]). They began asking themselves the question: Can we find a test site within the continental United States?

A European would smile at such a question about the home continent. But in the United States, there were still large tracts of unoccupied or very sparsely occupied land. I remember seeing Edward Teller, in 1950, poring over a table-top-sized map of the United States looking for likely test loca-

---

*The US stockpile of nuclear weapons declined from a high of more than 31,000 in 1967 to an estimated 7,700 in 2013, with a projected 3,600 in 2022.[8] It is reasonable to suppose that even as the numbers decline, newly designed weapons are replacing those of older design in the stockpile.

tions. He pointed to quite a few. (Why Teller was involved in site selection I don't know, but he liked to be near the center of things, and he very likely took the position that we *can* find and *will* find a continental site when some of his colleagues may have considered it a fruitless search. Teller was also part of the group that pointed to Aiken, South Carolina, at the meeting mentioned near the end of Chapter 9 (page 104). Why I was present at both meetings I don't know. I guess the curious were admitted.) In fact a site in Nevada some 65 miles northwest of Las Vegas was designated for nuclear weapons testing on January 11, 1951, and, with lightning speed, put to use. [11] Five tests took place there in Operation "Ranger" in the last week of January and first week of February 1951. In October and November of that year, in Operation "Buster Jangle" at the Nevada site, seven more bombs were exploded (one of them a fizzle). [9] It is very likely this Buster Jangle series that Carson Mark was working on in February 1951 when Stan Ulam barged into his office to tell him about a new idea.

Yucca Flat, part of the Nevada Test Site (as of 2011). The craters are places where the earth subsided after underground nuclear tests. *Courtesy of National Nuclear Security Administration.*

Over the years, the Nevada test site became the favored location for weapons tests. Of the 1,054 reported tests in the period 1945–1992, 928 were carried out in Nevada, 828 of them underground and 100 above ground.* The remaining tests were carried out mostly in the Pacific (106) with a scattering (20) in other places, including even two in Mississippi. [9]

Just three of those thousand-plus tests come into my story, all three conducted at Enewetak. They were Greenhouse George and Greenhouse Item in May 1951 and Ivy Mike on November 1, 1952 (October 31 in the United States).

By the time I arrived in Los Alamos in mid-1950, the authorities there had decided that thermonuclear burning needed some experiments, not just theorizing. In February of that year the "Family Committee" had come into being, with Edward Teller as its chair, to oversee thermonuclear developments. [14] (The name, I was told, reflected the fact that the committee was to consider a new family of weapons—or perhaps all families of weapons.) Members included leading lab scientists and engineers from various divisions—including those concerned with theory, experiment, chemistry, metallurgy, and testing. I was not a member but did attend some of its meetings, most of which were concerned with planning for the Greenhouse tests scheduled for the following May.

Available minutes of the Family Committee meetings run from its fourth meeting in March 1950 to its twenty-seventh meeting on November 15, 1950. [15] At that meeting, Norris Brad-

---

*Some of the above-ground tests in Nevada produced unacceptably high concentrations of radioactive fallout in locations such as St. George, Utah, where, beginning in the 1950s, elevated rates of cancer were reported. [12] There was also at least one serious incident of fallout from a test in the Pacific, where, in 1954, the crew of the "Lucky Dragon No. 5" fishing boat received high doses of radiation. The captain died seven months later and the lives of other crew members may have been shortened. [13]

bury, the lab director, reminded the committee that its charge was to look beyond Greenhouse at all phases of the thermonuclear program. [16] To what extent it did so I don't know since if any minutes were recorded for meetings after November 15, they remain shrouded in secrecy. To call the minutes of the earlier meetings "available" is a slight overstatement. They are works of modern art, studies in black and white, containing fully blacked out pages intermingled with partially whited out pages, with a few sentences peeking out here and there. The sentences that do make it into the light show that the committee was largely concerned with experiments planned for the George and Item "shots" at Greenhouse—experiments with names like FLUMEX, GANEX, TENEX, DINEX, and PHONEX. (A saying at the lab at the time was that if a really clean experiment were ever designed, it would be called KLEENEX.)

Part of my job in my first months at Los Alamos was to carry out calculations on the expected performance of some of these experiments. The George shot was the more complicated of the two. It used a large fission bomb, reportedly the largest to that date (225 kilotons), to ignite a small quantity of a liquefied (and frigid) mixture of deuterium and tritium. This was in the spirit of the 1946 invention of von Neumann and Fuchs that I described in the previous chapter—although, as I mentioned there, I don't recall the von Neumann-Fuchs invention ever being discussed in connection with planning for George. The ignition occurred because radiation from the fission bomb ran out ahead of the expanding matter from the fission explosion. The fission bomb was, according to descriptions now posted, cylindrical rather than spherical—perhaps the only one of its kind ever made. [17] That geometry meant that there was a channel out of which radiation could pour onto the nearby container of liquefied deuterium and tritium. Von Neumann and Fuchs had considered a gun-type weapon to achieve the same end. The questions of interest were: Did the DT mixture burn? If so, what fraction of it was consumed?

What temperature was reached in the DT container? (Energy release from the DT burning, even if complete, was inconsequential relative to the energy of the fission trigger.)

It was to answer these and related questions that all the diagnostic ...EX's were devised. It is a tribute to the skill of the experimental physicists and the engineers that so much could be learned in so short a time—much less than a thousandth of a second. The fast (*very* fast) cameras, the radiation detectors, the particle detectors: All had to send their reams of data, through wires and over the airwaves, in the instant before they breathed their last—true swan songs.

One of the most straightforward ways to verify DT burning is to look for the 14-MeV neutrons that are emitted (see the reaction equation on page 110). They stand out because they are more energetic than any other neutrons coming from other sources, and they hasten to the detectors faster than anything else except light. A neutron of this energy travels at about one-sixth the speed of light. Atoms of metal from the exploding bomb, even at their extreme temperature, are a few hundred times slower than that. Suppose that an array of detectors is located 30 meters (about 100 feet) from the bomb. Light will reach the detectors in a tenth of a microsecond, 14-MeV neutrons in six-tenths of a microsecond, and atoms of iron (from a 15-million-kelvin temperature maelstrom) in about 200 microseconds. The work of the detectors needs to be completed in something like one ten-thousandth of a second.

To everyone's relief, the unconventional fission bomb exploded as predicted, and the diagnostic experiments worked. The data from Greenhouse George (May 9, 1951) indicated that the DT mixture had been heated past its ignition temperature and had indeed burned—the first thermonuclear reaction to be achieved on Earth. There was even evidence that the DT container had experienced some radiation implosion and/or plasma-induced implosion of the kind von Neumann and Fuchs had visualized, although implosion had not been

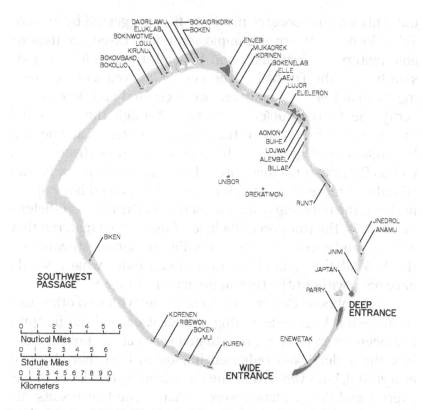

Enewetak atoll as of 1948. The island of Elugelab (here called Eluklab) is near the top, close to the atoll's northernmost point. *Courtesy of Defense Nuclear Agency.*

planned. George had been "put to bed" well before the ideas of radiation implosion and an equilibrium Super had surfaced [18] and seemingly after the work of von Neumann and Fuchs had receded from the hippocampi of the various physicists.

Greenhouse Item (May 25), like George, tested new thermonuclear strategies, but with less complexity and a higher degree of confidence that all would work as planned. Item consisted of a spherically imploded fission bomb (not unlike the ones tested at Alamogordo and dropped on Nagasaki) with a small ball of deuterium and tritium at its center—this time the DT mixture was a gas at ordinary temperature, not a liquid. This was the booster principle first suggested by Edward Teller in 1945.* When the implosion compressed the fissionable material to criticality and beyond, the resulting explosion heated the DT ball at the center (as well as compressing it). Again 14-MeV neutrons were created and detected to verify the thermonuclear burning. Although the DT added only a small bit directly to the energy released, it augmented the fission yield greatly by furnishing neutrons (from DD as well as DT reactions) that entered the fissionable material and stimulated more fissions before everything could blow apart. In short, the boosting principle increased the bomb's efficiency—that is, the fraction of its load of fissionable material that actually underwent fission. According to some estimates, [20] the 45.5-kiloton yield of Item was about twice what it would have been without the thermonuclear assistance.

Greenhouse George and Item (along with two other unrelated—but large—tests, Dog and Easy, carried out in April) had been set in motion more than a year earlier, not only before the Teller-Ulam radiation-implosion idea that changed everything, but even before the discouraging results of Ulam-Everett and Fermi-Planck were in hand. The test results, although hardly irrelevant—they provided lots of useful data and validated new designs—were not critical to deciding in what directions to go next. The equilibrium Super was on the front burner.

In the same month that the George and Item results flowed into Los Alamos and were being analyzed, I was in Albuquerque buying a camera, a small motorcycle, and goggles, in preparation for my return to Princeton.

---

*In fact, Teller applied for a patent on the booster idea before he left Los Alamos at the end of World War II. [19]

Chapter 11
Constructing Matterhorn

When John Wheeler interrupted his stay in Paris to join the H-bomb effort at Los Alamos in early 1950, he imagined that he might need to remain at the Los Alamos lab for up to two or two-and-a-half years. There was, at the time, no clear path to success on the bomb—just the likelihood of a lot of hard work to bludgeon nature into submission.

His attitude changed after he had been in Los Alamos for half a year or so—not about the need for hard work, only about where that work should be performed. There were several reasons. First, he and Edward Teller had been notably unsuccessful in attracting other top scientists to the work. To be sure, Fermi, von Neumann, and Bethe, who were the best of the best, agreed to continue as consultants, as did the very young and very brilliant Dick Garwin. One physics professor, Lothar Nordheim, from Duke University, signed on for two years. That was it. No other fish bit, although the lures were cast far and wide to researchers at top universities. (Teller and Wheeler personally, not the lab authorities, were apparently the recruiters-in-chief, although Teller may have induced Norris Bradbury, the lab director, to sign some of the letters.) Some of those contacted didn't see the need for a "crash program." Others thought that their work on pure physics and with a new crop of graduate students was more important than weapons work. Others felt that they had done their bit during the recent war and were not ready to go back on what would be effectively a war footing. Neither the Soviet threat, which so alarmed Teller and Wheeler, nor the scientific challenge were enough to pry loose top physicists from their postwar pursuits. These are the themes of responses as described by John Wheeler to John Toll and me.

Wheeler began to wonder if geography was the problem. Could it be that if some piece of the work were done at a leading research university (Princeton) top physicists would join the effort? (It was, and they didn't. Wheeler's unsuccessful recruitment effort is something I address later in this chapter and again in Chapter 15.)

When, in the late fall of 1950, Wheeler made his pitch to the Los Alamos management for an ancillary effort to be based in Princeton, he added another argument, that important calculations could conveniently be carried out at the still-under-construction MANIAC computer at Princeton's Institute for Advanced Study. (Again, hope and reality diverged. The MANIAC, in 1951 and 1952, was unable to deliver reliable performance, and Wheeler's Princeton group turned to other computing resources.) Whether Wheeler attached real importance to being in the MANIAC's backyard is doubtful. But when you are making a sales pitch, you marshal whatever arguments are available.

As important as any other consideration but no doubt missing from the official records was the attitude of Wheeler's wife Janette. After her April 1950 arrival in Los Alamos, she tried hard to make her Bathtub Row house into a home and to fit into the life of what she perceived as a company town. She made friends with Francoise Ulam and continued her friendship with Mici Teller but remained bothered by the apparent correlation of town social life and rank in the lab. In truth, she desperately missed Princeton. In the fall, adding to her other discontents, she reached the conclusion that the Los Alamos public schools were substandard. I recall hearing Janette and John Wheeler discussing the possibility of sending their two older children (then twelve and fourteen) to a boarding school in Switzerland if John remained another year in Los Alamos.

(Their youngest child, Alison, then eight, told me in 2013 that she enjoyed her year in Los Alamos but learned nothing. When she did return to Princeton, in 1951, it was to enroll in Miss Fine's School, with its high standards matching those of her school in France.)

Driven by the various motivations, Wheeler prevailed. His first approach on paper to a Princeton official came in a letter of January 8, 1951 addressed to Allen Shenstone, his department chair (whose tenure in that position was to expire at the end of the month). [1] In his last two days in office, Shenstone made clear his opposition to the idea in communications to university president Harold Dodds and to Harry Smyth, his former Princeton colleague now in Washington as an Atomic Energy Commissioner—but to no avail. [2] [3] Within a few weeks, Smyth had expressed guarded support and the new department chair, Don Hamilton, relayed Smyth's support and his own (with caveats concerning where and how secret work might be conducted) to President Dodds. [4] By April 1951, Project Matterhorn existed on paper (although not yet with that name), and Hamilton was communicating with Wheeler on the salary structure for the new project. [5] By June, project staff were at work in Princeton.

That Princeton embraced a military research project at all is astonishing, and that it happened so swiftly is a seeming miracle. Here was one stubborn scientist (Wheeler), backed up by another stubborn scientist (Teller), arrayed against the massive inertial blocks of the Los Alamos lab; its patron, the Atomic Energy Commission (AEC); and Princeton University, with its policy against secret work. (Fortunately, the United States Congress wasn't involved. The AEC seemed to have all the money it needed and then some. The University of California, another inertial block, also played no role except pro forma when there was a contract to be signed.) Going from the expression of an idea to the reality of a satellite lab took less than six months.

How did it happen? Mostly, I conclude, because of the tenor of the times. One after another, events of 1948–1950 brought home the reality of the Cold War and stoked fears of a hot war. The Berlin Blockade (1948–1949) put an end to any Soviet-American partnership and seemed to make real the "Iron Curtain" cited by Winston Churchill in 1946. The first Soviet nuclear explosion (August 1949, announced in September) ratcheted up fear of war, and led eventually to children being asked to practice cowering under their desks. Truman's January 1950 statement, widely interpreted as authorizing a crash program to build an H Bomb, added to the nervousness. When Klaus Fuchs was arrested in February 1950 for transmitting nuclear secrets to the Soviets (he had confessed in January), Americans felt even more vulnerable. Then came the Korean War in June and Chinese intervention in it in October, leading to President Truman's declaration of a National Emergency on December 16, 1950. [6] (In Don Hamilton's February 1951 memo to President Dodds supporting the Wheeler proposal, [4] he cited the "present state of emergency," expressing a perspective that was showing up in other university memoranda of the period.)

At the same time, our elected officials in Washington and in State Houses around the country were doing their part to foment an almost hysterical anticommunism, feeding fear that our entities of government, federal and state, were being infiltrated under guidance from Moscow. The worst of the lot was Senator Joe McCarthy, who in February 1950 brandished a list of supposed Communists in the State Department, and went on to ride the nation's paranoia right through an easy reelection in Wisconsin in 1952, until his final censure by his Senate colleagues in 1954. State governments did their part by requiring loyalty oaths such as the ones I encountered in California (actually in New Mexico) and Massachusetts. In Washington, the House of Representatives, through its Un-American Activities Committee (HUAC), went on a witch hunt starting with

Hollywood personalities in 1947 and progressing to physicists in 1949.

Because the anticommunist fervor of the times was so significant in making Wheeler's Princeton project possible, I address its effects on academia and, in particular, on one physicist, in the next chapter.

While working at the University of Chicago's Met Lab in World War II, as I mentioned in Chapter 8, John Wheeler often declined invitations from his fellow scientists to go out to lunch, so that he could sit at his desk with a sandwich, doing his "Princeton physics." [7] This dogged commitment to pure physics remained with him in Los Alamos and at Project Matterhorn. It was what accounted for the three desks he wangled from the Zia Company for his Bathtub Row home in 1950. And accounted for his delight when, in May 1952, in the midst of managing Matterhorn and helping to plan for the "Mike" test, he gained approval to introduce and teach a new course on relativity that fall. I recall seeing Wheeler in September 1952 with a copy of Peter Bergmann's textbook on relativity in one hand and a sheaf of computer printouts on H-bomb calculations in the other. (Later, with two of his former students, Wheeler wrote what was for many years the definitive textbook on the subject, *Gravitation.* [8])

It was not surprising, then, when Wheeler decided in the fall of 1950 that he and his two students at the time, John Toll and I, should submit papers to be presented at the December 1950 meeting of the American Physical Society in Los Angeles. APS then held its biggest meeting of the year in New York in late January (at that time a semester break at many universities)—understandably known simply as the New York Meeting. Preceding it by a month each year was a west-coast meeting held in the week between Christmas and New Year's, alternating between northern and southern California. In 1950 UCLA was the host. (I should add that simultaneous sessions were then a thing of the future. One lecture hall sufficed for the sequential sessions that December.)

So, in the fall of 1950, in went three abstracts, two by Wheeler and Toll (one talk to be delivered by each of them) and one by Wheeler and me, with me as the presenter. The Wheeler-Toll papers grew out of Toll's dissertation in progress, and dealt with topics as esoteric as the deflection of one photon of light by another.* The Wheeler-Ford paper was a quantum extension of classical work that I had done two years earlier in Wheeler's course, and dealt with what we called the "quite possibly mythological entity," the magnetic monopole. In short, our papers had nothing to do with thermonuclear burning. [9]

Teller decided to join the party, although, as far as I can determine, without submitting an abstract or delivering a paper. He may have just wanted to recharge his intellectual batteries; or to spend more time talking privately with Wheeler about H-bomb prospects; or simply to take a break. He and Wheeler decided to take their wives along to Los Angeles and make a bit of a vacation out of the trip (or the wives decided to come along). They also wanted to see more of New Mexico, Arizona, and California than could be glimpsed from a train window. But they didn't feel they could spare the time to drive in both directions. That is where Toll and I entered the picture. We junior members of the team would drive the wives to Los Angeles while the husbands took the train and talked about thermonuclear reactions. The husbands and wives would drive back to Los Alamos together while Toll and I would ride back by train. It worked out as planned. What I remember most about the return trip was that we boarded the train in Pasadena on New Year's day after watching the Rose Bowl Parade.

---

*Not so esoteric now, having been finally observed and measured in 1997.

What I remember most about the trip out was the cold. Janette Wheeler, Mici Teller, John Toll, and I decided to make it a two-day trip, breaking midway by camping out at the Grand Canyon. We turned north from Flagstaff well after dark on the first day and had no trouble at all finding a deserted spot on the high mesa just south of the Canyon. There was nothing else but deserted spots. We got out our sleeping bags, undressed, got into the bags, and had a good night's sleep. Getting going in the morning was a little more challenging. The temperature was surely below zero. The few minutes it took to get out of the sleeping bags, into our clothes, and into the Wheelers' Chevrolet seemed agonizingly long. Fortunately, a lodge with a roaring fire in its fireplace was not far distant. Never have I enjoyed a breakfast more.

Janette Wheeler, early 1960s. *Courtesy of Alison Wheeler Lahnston.*

Mici Teller, 1950s (cropped from an original that included Edward Teller). *Photograph by Jon Brenneis, courtesy of The Bancroft Library, University of California, Berkeley (and with thanks to Istvan Hargittai and Wendy Teller).*

This was my first trip to California, so I was wide-eyed at the three-car garages in Beverly Hills and the sizes of the salads and the pepper grinders in the restaurants. It was, for both John Toll and me, also our first presentation at a Physical Society meeting. So we rehearsed our ten-minute talks with an audience of one in the hotel room we shared.

By the time of this trip to Los Angeles, Wheeler had secured the blessing of the lab director, Norris Bradbury, for the Princeton project, provided it was clearly an extension of the Los Alamos lab and answerable to Los Alamos, a condition perfectly acceptable to Wheeler (although the project staff would be Princeton employees). [10] Wheeler was not yet in a position to invite others to join the project, except for Toll and me. However, as early as the summer of 1950, he had raised the possibility of a Princeton H-bomb project with his friend Lyman Spitzer, the Princeton astrophysicist, and had secured Spitzer's agreement to help. In December 1950, after an observing session at the 100-inch Mt. Wilson telescope in California, Spitzer stopped by at Los Alamos to be briefed on thermonuclear progress. [11]

Two or three months later, in February or March 1951, following a ski holiday in Aspen, Colorado, Spitzer visited Los Alamos again, this time fired up about an idea for controlled fusion power that he had formulated while skiing. [12] He was

Lyman Spitzer, unknown date.
Photograph by Ulli Steltzer, courtesy of AIP Emilio Segrè Visual Archives.

ready to propose to Wheeler that if the Princeton project became reality, it could pursue both controlled and uncontrolled thermonuclear reactions—for electric power and for bombs.

Almost from the day of its discovery, nuclear *fission* was recognized as a likely source of controlled power, not just a source of violent explosions. Over the years, controlled power (reactors) and uncontrolled power (bombs) were developed in parallel—although largely by different groups. *Fusion* followed a different path—or paths. Very early, such as at the time of Oppenheimer's 1942 gathering in Berkeley, physicists recognized fusion's potential for explosive energy release—a goal achieved a decade later. Controlled fusion power appeared to be a more formidable task—and so it has proved to be. Yet by 1951 there was a sense of guarded optimism that it was a goal that might be reached in a decade or so. New ideas, especially if advanced by someone of Spitzer's stature, were worth listening to.

On the morning of Spitzer's planned visit to the lab, some security clearance glitch delayed his admission, and he was stewing about it. So John Toll and I were dispatched to the Los Alamos Lodge to have lunch with Spitzer and calm him down. Over lunch he shared his idea with us. Take a torus (a slender doughnut), fill it with a hot (*very* hot) plasma of deuterium, grab the torus with two hands, and twist in opposite directions until the torus becomes a figure eight. Thread a magnetic field through the device. Then, Spitzer reasoned, the deuterium nuclei, instead of drifting in one direction and hitting a wall, would drift back and forth within the enclosure, avoiding collision with the wall long enough to permit thermonuclear "burning" to take place.

Later in the day, the security problem was resolved and Spitzer was able to unveil his idea to Wheeler—and no doubt to Teller and others. Either that very day, or perhaps soon after, Wheeler and Spitzer put their heads together and decided to make the Princeton project a double-bodied (and two-headed) enterprise, to pursue controlled and uncontrolled fusion side by side. Both aspects of fusion energy were, at the time, classified secret. Within a few years controlled fusion research began to open up (especially after the 1955 Atoms for Peace conference in Geneva), and by now is entirely unclassified. Yet in December 1951, with Matterhorn just half a year old, an official of the New York Operations Office of the Atomic Energy Commission felt compelled to write to the financial officer Roy Woodrow at Princeton reminding him that personnel in the two branches of Matterhorn were not permitted to discuss their respective work with each other. [13] This was a stricture to which we paid no attention.

Despite this agreement between Wheeler and Spitzer, as late as April 1951 memoranda among officials at Princeton were referring to "Wheeler's project." [14] Spitzer was, in fact, named to an oversight committee, and not replaced on that committee until May. [15]

By early March 1951 Wheeler was feeling confident enough about the reality of his Princeton project that he felt free to write to notable physicists around the country inviting their participation. Here is language from a letter he wrote to Harvard Professor Wendell Furry on April 6: [16] "I hope conditions will permit [your continued work in pure physics]. They may not. My personal guess is an appreciable chance of war by September. You undoubtedly have your own probability estimate. You may be doing some thinking about what you will do if the emergency becomes acute." Wheeler wrote in a similar (or identical) vein to others. In his autobiography he says that he contacted 120 senior colleagues by letters, telephone calls, telegrams, and personal visits, among whom only one, Louis Henyey, an astrophysics professor at UC Berkeley, was sufficiently moved to join the project. [17] "So I turned to bright, interested younger people," Wheeler wrote, and added, "Fortunately, the mental muscle of a theoretical physicist can reach peak performance early. In this mostly under-thirty group we had a lot of talent, enough to get the job done." [18]

Up until May 1951, "Wheeler's project" had no name. That is uncharacteristic of Wheeler, famous for his coinages and for his flair. The under-the-radar namelessness was resolved in a meeting among Wheeler, Spitzer, and a senior financial officer of the University, Roy Woodrow, in a Palmer Lab office on May 25.* [19] (It must have been a little unusual for Woodrow to journey across campus to meet in the physics building. When I later sat in on a meeting with these three, it was in Woodrow's office.) In capsule form, here is the conversation that apparently took place on that occasion. Woodrow: Your project needs a name. Spitzer: How about Matterhorn? That symbolizes challenge and also reflects the fact that I had an idea for a device while skiing. Wheeler: OK, Lyman, you can name the project if I get to name your device. How about Stellarator? That symbolizes harnessing the energy of the stars.

Matterhorn, Zermatt, Pennine Alps, Switzerland. *Courtesy of Freepik.*

*By chance the Greenhouse Item shot was fired at Enewetak on this same date (on the other side of the international date line—May 24 in Princeton).

Many years later, Spitzer wrote that he suggested the name "since the Matterhorn is a spectacular peak which can nevertheless be climbed with sufficient effort." [20]

I don't know if they shook hands. But that was that. Matterhorn B (for bomb) and Matterhorn S (for Stellarator) officially came into existence that month and existed in tandem until Matterhorn B shut down in 1953. Then, in 1958, Matterhorn S changed its name to the Princeton Plasma Physics Laboratory at the time that research on controlled fusion was released from secrecy restrictions. The Stellarator has evolved and remains an object of study, not just in Princeton but in laboratories around the world. As of this writing, the international newsletter "Stellarator News" is published regularly by Oak Ridge National Laboratory "by and for the stellarator community." [21]

Another thing that Princeton's administration and Wheeler and Spitzer readily agreed on was location. Matterhorn would be located at the University's newly acquired 825-acre Forrestal Research Center* several miles from the main campus. This location gave comfort to those at the university who wanted an insulated layer between academia and secret projects, and it was no hardship to Matterhorn. (The University had purchased this property from the Rockefeller Institute for Medical Research when that institute consolidated its activities in New York City in the spring of 1951 preparatory to becoming Rockefeller University. Reportedly, as part of the sales agreement, Laurance Rockefeller (Princeton '32) insisted that the University's new acquisition be named after James Forrestal (Princeton '15), the nation's first Secretary of Defense.) [22]

---

*Now the Forrestal Campus, with double the initial acreage.

I had been working diligently at Los Alamos and expected to do the same in Princeton. John Wheeler had no objection to my taking a two-week break between jobs—literally between. I decided to ride a motorcycle from one place to the other, circuitously through Utah, Wyoming, Montana, the Dakotas, and Wisconsin. In mid-May 1951 I hitched a ride to Albuquerque and came back to Los Alamos on a BSA 125-cc motorcycle. It had one cylinder and a two-gallon gas tank (in which the gasoline needed to be mixed with oil), delivered 4 horsepower, and was understandably called the Bantam. It was available in one color, mist green, and was capable of speeds in excess of 50 miles per hour downhill.

Without waiting for the results of Greenhouse Item to come in, I parked my Carryall and most of my belongings with a friend, and took off. In Utah, I raced a freight train, barely pulling ahead. In Wyoming, I flagged down a motorist in a torrential downpour, asking him to carry my camera to the lodge near Old Faithful. Camera and I survived the soaking. In Madison, Wisconsin, on a perfect spring day, I joined a group of university students for a picnic on a grassy slope overlooking a lake. In Cleveland, Ohio, I stopped to visit a girlfriend named Libby. She was enchanted (I think) with my raccoon-like appearance. I had large white circles around my eyes where my goggles had blocked the sun. Her father, however, was alarmed when I told of my back pain that got worse every afternoon. He was an executive used to exercising authority. Nothing I said could deter him from crating the Bantam for shipment to Princeton and sending me on by train.

I arrived refreshed in Princeton, where the motorcycle served briefly as my principal means of transportation (augmented by a bicycle and, before long, by a British roadster—I just *had* to spend all that money I had made in Los Alamos). I can't remember at all how my belongings got to Princeton. I suspect that John Toll brought them in his car.

Chapter 12

Academia Cowers

Among the people caught up in HUAC's net was David Bohm, an assistant professor of physics at Princeton. (Another was J. Robert Oppenheimer's brother Frank—also a physicist—who, after being forced out of a professorship at the University of Minnesota, became a cattle rancher and high-school teacher in Colorado, and later, to the benefit of millions, director of San Francisco's Exploratorium.* [1]) I discuss Bohm here because his case had some influence on Princeton's faculty, pro and con, and thus some bearing on Princeton's receptivity

Frank Oppenheimer when he was a physics teacher in Pagosa Springs, Colorado, late 1950s. *Photograph © Stanley Fowler, courtesy of the Exploratorium.*

---

*Another was G. Rossi (Ross) Lomanitz, who was discharged from Fisk University after coming afoul of HUAC in 1949.[2] Like Bohm, he was part of the Berkeley group around Robert Oppenheimer in the early 1940s. He finally found a receptive professional home at New Mexico Institute of Mining and Technology, where he and I were colleagues in the late 1970s and early 1980s.

to Wheeler's proposal. Also I had a personal link to Bohm. He was co-author of my first published paper and was, for a time, my roommate.

Not long after the end of World War II—probably in 1946—John Wheeler became aware that David Bohm, then twenty-eight, might be an especially promising young theoretical physicist.[3] Bohm was in Berkeley, where he had worked throughout the war, guided by—or at least inspired by—Robert Oppenheimer. Some of his wartime projects were related to the atomic-bomb effort, some were in pure physics. (He was barred, reportedly by General Groves himself, from actually moving to Los Alamos.[4]) Wheeler was especially intrigued by the fact that Bohm had shown an interest in the fundamentals of quantum mechanics. (Wheeler was, at the time, working on more practical problems related to atomic nuclei and elementary particles, but he retained throughout his life an itch to get deeper into whatever principles might underlie quantum mechanics.) After interviewing Bohm in Berkeley, Wheeler persuaded his Princeton colleagues to offer Bohm a job as an assistant professor—a so-called tenure-track posi-

David Bohm, 1949. *Library of Congress, New York World-Telegram and Sun Collection, courtesy of AIP Emilio Segrè Visual Archives.*

tion. Bohm accepted and arrived to take up his duties in Princeton in January or February 1947. [5]

As it turned out, Bohm's research interests at Princeton centered on plasma physics—a field in which he did quite notable work. Despite the quality of the work, the new direction was a disappointment to Wheeler, who had been looking forward to intellectual discourse with Bohm. Wheeler later admitted that he did not resonate with Bohm's style, [6] and he had no sympathy for Bohm's left-wing politics or his refusal to cooperate with HUAC. So when Bohm was arrested and charged with contempt of Congress in 1950, Wheeler was not among his Princeton defenders.

In 1948–49, my first year at Princeton, I took a year-long course in quantum mechanics with Bohm—a superb course with about a dozen students that was fun as well as instructive. In keeping with Bohm's preference for working late and sleeping late, the course was taught in the evening.

Toward the end of that academic year, in May 1949, Bohm was called to testify before HUAC.* He sensed that the committee wasn't so much interested in him as in his Berkeley friends, whom the committee hoped to paint as dangerous Communists. Bohm invoked the fifth amendment and declined to testify. We graduate students had no respect for HUAC and a great deal of respect—even affection—for Bohm, so there was no question where our sympathies lay. Most of his faculty colleagues were probably supportive, too, although not all.

An entire year-an-a-half elapsed with no repercussions for Bohm, so campus life went on pretty much as before, although we were aware of other physicists who were being forced out of their jobs for their supposed Communist ties. It wasn't until December 1950, by which time I was in Los Alamos, that Bohm was arrested and indicted for contempt

---

*He was subpoenaed on April 21, 1949 and testified five weeks later on May 25. [7]

of Congress.[8] Princeton University wasted no time reacting. President Dodds at once issued a statement: "Dr. David Joseph Bohm, assistant professor of physics, has been suspended from all teaching and other duties at Princeton University." [9] Bohm was paid through June 1951, when his appointment ended, but was barred from campus, or at least from Palmer Lab, where the physics department was housed. Students who wanted to talk to him had to meet him off campus. Much later, in a 1986 interview, Bohm had this to say about that six-month period. "I made much more rapid progress during that time than at any time before." [10]

In May 1951, the very month in which Project Matterhorn was being established in Princeton, Bohm went to trial and was acquitted. Despite efforts on his behalf by some of his faculty colleagues, he was not reappointed at Princeton. Finding no other jobs in the United States, Bohm went to Brazil, and later to Birkbeck College in London, where he remained until his death at the age of seventy-four. Wheeler's initial hope was, in the end, realized. Bohm devoted a great deal of his professional activity in his later years to exploring the fundamentals of quantum theory.

The kinds of problems faced by Bohm, Frank Oppenheimer, and Ross Lomanitz were not the only fallout from the "Red scare" of the late 1940s. Security clearance was instituted as a requirement for employment in numerous Federal agencies as well as in university projects to which the money flowed from these agencies. Some scientists, as well as technicians and support staff, were denied clearance—and, in many cases, thereby denied employment—on the basis of unsubstantiated charges. For agencies faced with thousands of cases, it was no doubt often easier to deny clearance than to explore charges with care. To help address this problem, the Princeton astro-

physicist Lyman Spitzer—later to head Project Matterhorn S—
and a few of his colleagues formed the Scientists Committee
on Loyalty Problems (SCLP) in the fall of 1948,* around the time
that I was beginning my graduate study at Princeton. This became a committee of the Federation of American Scientists[†]
and soon acquired several score notable scientists around the
country as sponsors and consultants. [12] Over the three years
of its existence, the committee tried to help individuals with
clearance problems and also tried to influence procedures
of the Federal agencies. Somehow I became a member of the
committee and found my name emblazoned on the committee's letterhead. I have no recollection of how this happened.
I suspect that I was a sympathetic student willing to help keep
records and oversee correspondence. When my own Q clearance was granted in about one month in 1950, I realized that
I must have a record perceived as very clean—despite my
having participated in the summer of 1949 in an international
gathering of students in Germany at which there was a good
deal of left-leaning talk, and despite my friendship with David
Bohm.

Bohm was, in fact, a member of SCLP during its first year
of existence. Shortly after his HUAC appearance, in late May
1949, he offered to resign from SCLP. [13] The senior members of the committee were torn. They wanted to support his
stand, but they also felt that his membership on the committee would weaken the committee's clout, such as it was, with
government agencies. At a meeting of June 4, they (or perhaps
we—I can't now remember if I was present at this meeting)
resolved the matter by resigning en masse, and so dissolving
the committee. [14] But they (we) did so with the understanding
that SCLP would quickly be reconstituted, with new members,
not including Bohm and not including certain members who

---

*Its first meeting was on September 25, 1948. [11]

[†]Founded in 1945 as the Federation of Atomic Scientists.

were based far from Princeton and had not been active in the committee's deliberations and actions. So the dissolution and reconstitution took place at once. The committee then continued its activities on behalf of fair security clearance procedures for two more years. It went out of business in July 1951. [15] (In June 1950, a year after the mass resignation, Sam Goudsmit,* a physicist at Brookhaven National Lab on Long Island, submitted his resignation from the committee. I have not been able to determine why. Perhaps he felt that, being at some distance from Princeton, he wasn't able to contribute much. In any case his resignation was not accepted. [16])

Some time after his testimony but before his indictment, I happened to be chatting with Bohm after we had both listened to a talk by a Princeton mathematician on antinomies (paradoxes). Bohm smiled and said, "Congress should appoint a committee charged to investigate all those committees that do not investigate themselves."

Around the same time (1949–1950), I mentioned to Bohm that I was interested in leaving the Graduate College and moving into quarters in the town of Princeton. He told me that he had a large room containing two beds in a house on Nassau Street and that I was welcome to join him there, which would save us both some money. So for several months I was David Bohm's roommate. As it happened, there was very little beneficial "data transfer" from his brain to mine. When he came in and went to bed around 3:00 a.m., I was long since asleep, and when I left around 8:00 a.m., he was still sleeping soundly.

But we did have some very fruitful scientific interaction. In the spring of 1950 I was asked to make a presentation to the physics department's "Journal Club." This was an evening

---

*Goudsmit is best known to the public for his leadership of the Alsos mission, which, in late 1944, followed troops into France and Germany and discovered that German scientists had made little progress toward an atomic bomb. He is best known to physicists as the co-discoverer of electron spin and the long-time editor of journals of the American Physical Society.

gathering every week or so at which faculty and graduate students presented summaries of recently published articles by non-Princeton physicists. Typically there would be three twenty-minute presentations in a session. I was to report on a just-published paper by several experimental physicists at Oak Ridge. They had measured the strength of interaction of very low energy neutrons, or slow neutrons, as they were called, with numerous atomic nuclei. Not knowing anything else to do, I looked for some pattern in the published data by recalling a technique I had learned in Bohm's quantum-mechanics course. I treated the nuclei as transparent spheres—like little glass orbs, so to speak, rather than like hard billiard balls—and found that the data were indeed consistent with that model, which included appropriate growth in the size of nuclei as one marched up the periodic table. The assembled faculty showed more interest than I would have expected. After the session, Bohm came up to me and said, "Ken, what you have done may be quite significant. I can work with you to prepare a paper for publication." (At the time, physicists were imagining nuclei to behave more like opaque, not transparent, spheres. For reasons other than my small piece of work, that was about to change as the "nuclear shell model" came to be accepted.)

I was more than happy to accept Bohm's offer. Within days—or at most a couple of weeks—we had prepared and submitted a short paper with the title "Nuclear Size Resonances." It was published as a "Letter" in *Physical Review* on August 15, 1950, by which time I was hard at work in Los Alamos. [17]

When I returned to Princeton the next May or June, I did not see Bohm. His non-reappointment was by then old news, and he may have left town by then. Within the physics department, collegiality won out over hard feelings. If there was residual tension between those who supported Bohm and those who didn't, it wasn't evident to this observer. Wheeler never spoke to me of Bohm's politics or his run-in with HUAC. He just told me that Bohm had not performed up to the department's expectations. So, like many another non-tenured assistant professor before and since, he was not promoted or reappointed.*

---

*A year earlier, in 1950, Princeton had decided not to retain Robert Hofstadter as a physics faculty member. He moved to Stanford University and in 1961 won the Nobel Prize for his work there on a linear accelerator. Failing to win tenure at Princeton is the best thing that ever happened to Hofstadter.

Chapter 13
New Mexico, New York, and New Jersey

Simplicity marked Matterhorn's beginnings. For office space we were assigned a metal shack that had held experimental animals for the Rockefeller Institute, and the aroma lingered. For the first week that we were there, John Wheeler, John Toll, and I slept on cots in a cavernous boiler room of the former Institute. Wheeler was waiting for tenants to vacate his Battle Road house. Toll and I were looking for places to live in town. Fortunately, the boiler room's amenities included a bathroom and shower facilities.

The (somewhat aromatic) shack in which Project Matterhorn got its start in 1951.
Photograph by John Peoples, courtesy of Princeton University.

Understandably, for a few weeks organizing the project took as much time as doing physics. Wheeler led the recruitment of young physicists to augment Toll and me. Toll served as Wheeler's general deputy (his "strong right arm" [1]). Toll's "people skills" and talent for administration were already evident. As I recall, Toll also arranged for security to be put in place. That meant locked filing cabinets and round-the-clock guards. I was assigned the task of finding secretaries and human "computers." Because I had never done anything of the kind before and was intrigued by new challenges, I plunged into that task and quickly found and hired half a dozen well qualified young people—all female, as it turned out. The picture on this page shows most of the Matterhorn B team in 1952, support staff in front, scientists behind. The gender imbalance is evident.

Most of the Matterhorn B team in 1952. Front row, left to right: Margaret Fellows, Peggy Murray, Dorothea Reiffel, Audrey Ojala, Christine Shack, Roberta Casey. Second row: Walter Aron (with mountain-climbing rope), William Clendenin, Solomon Bochner, John Toll, John Wheeler, Ken Ford. Third and fourth rows: David Layzer, Lawrence Wilets, David Carter, Edward Frieman, Jay Berger, John McIntosh, Ralph Pennington, unidentified, Robert Goess. *Photograph by Howard Schrader, courtesy of Lawrence Wilets estate.*

I was also dispatched to the Princeton Post Office to get a Post Office Box number for Matterhorn (B and S combined). We were assigned P. O. Box 451. If you go now to the Web site of the Princeton Plasma Physics Laboratory, you will find, more than half a century later, that its mailing address is P. O. Box 451.

When Matterhorn was conceived, there was no Teller-Ulam idea. A few months later, when Matterhorn set up shop and got to work, the Teller-Ulam idea of radiation implosion was the clear choice for the path ahead. Yet, in May and June 1951, there was no specific, well-defined design for the device to be tested the next year. We just had the general idea that a cylinder of thermonuclear fuel—either deuterium or lithium-6 deuteride—would be compressed and would, we hoped, ignite and burn. Even without a design, we could also assume that the cylinder holding the fuel would be made of uranium—ordinary uranium, consisting largely of U238, not the rare, expensive isotope U235 used in some fission bombs (I say "some" fission bombs because plutonium-239 was—and is—also a common component of fission bombs). The reason for this assumed choice of container was that neutrons generated in thermonuclear burning—especially the 14-MeV neutrons created in the DT reaction (see page 110)—are energetic enough to cause even U238 to undergo fission, so the container itself would be another "bomb," adding greatly to the energy released. Uranium was therefore the clearly preferred choice over steel or any other material to house the thermonuclear fuel.

Also, even without a specific design, there was an obvious way to divide up the work. The device would have to consist of a fission bomb to provide the initial jolt of radiation, a cylinder of thermonuclear fuel to provide the fusion energy (and a lot more neutrons), and a connecting structure and radiation channel between the two. In short, a three-stage device: a fission component, a link, and a fusion component (with a good deal of fission also showing up in that third component). The scientists at Los Alamos, with already half a dozen years of experience in designing fission bombs of various sizes and types, were clearly better qualified than any of us at Matterhorn to work on the fission bomb part of the design. They, along with their engineer colleagues at Los Alamos, were also better able to deal with the connecting link and radiation channel, however that might turn out to be configured. Wheeler and Toll and I, during our time in Los Alamos, had worked almost exclusively on thermonuclear burning—mostly in the context of the classical Super, but also specifically on the George and Item tests at Greenhouse (fired in May 1951). For that reason and because of the small size of the Matterhorn group, the Princeton end of the theoretical physics axis was logically assigned responsibility for calculations and design bearing on the fusion end of the device.

Of course, in reality, neither the work nor the device itself could be so neatly compartmentalized. It was a single, interconnected device. And it was a single, interconnected team. During the course of the 1951-52 year, we in Princeton kept in close, regular touch with our Los Alamos colleagues and made some trips to meet with them in person. (I remember one such trip when Wheeler telephoned Pennsylvania Railroad headquarters in New York City and explained why it would be in the national interest for the Broadway Limited to make an unscheduled stop in Princeton Junction to pick up some Matterhorn scientists bound for New Mexico. When that train did stop for us, the conductor, assuming that nothing less than college athletics could account for such an unusual event, asked if we were the Princeton basketball team and John Wheeler was our coach. A few of us were of above average height.)

Since the trip from Princeton to Lamy (the station serving Santa Fe) entailed a change of trains in Chicago,* it afforded the opportunity for us to meet with Edward Teller, which
we did at least once in the fall of 1951. He had decamped from
Los Alamos in September, returning to his professorship at the
University of Chicago, irritated with Los Alamos management
but quite eager to be briefed on what we were up to and to
make suggestions.

From the summer of 1950 through the summer of 1952,
first from a Los Alamos base, then from a Princeton base, much
of my own effort was devoted to programming and computing nuclear reactions and nuclear energy release—from both
fusion reactions and secondary fission reactions. At the beginning of that two-year period, the resources I had available
were a slide rule, a desk calculator, and Los Alamos's human
"computers." Soon added were IBM card-programmed calculators (CPCs), which I used in Los Alamos, at Sandia Laboratories in Albuquerque, and in an IBM building in New York City.
At the end of that two-year period, I was working with the
SEAC computer at the National Bureau of Standards in Washington, D.C. In the summer of 1952, the SEAC (Standards Eastern Automatic Computer) was probably the finest computer in
the world—although its memory capacity and its speed fell far
short of what is today available in the dullest of smart phones.
The SEAC's granddaddy, the ENIAC, was by then obsolete. The
SEAC's immediate progenitor, the MANIAC at the Institute of
Advanced Study, was in principle superior but in 1952 was still
too error-prone to compete seriously with the SEAC. In Chapter 15 I discuss my many all-night shifts on the SEAC.

---

*Although not a change of car. Our sleeper was disconnected from the Penn-
sylvania Railroad Broadway Limited and connected to the Santa Fe Chief or
Super Chief for departure a few hours later.

Most of our early calculating was on so-called "one-dimensional" configurations. Although the device was, of course, three-dimensional, limitations of computing power forced us to track changes of conditions along only one dimension, either "axially" or "radially." We sought to learn if a charge of thermonuclear fuel, heated to a certain temperature and squeezed to a certain density and pressure, could "burn" steadily down the length of a cylinder after being ignited at one end, or could burn outward from its center toward its encircling container without fizzling. One needed to keep track of a set of variable quantities such as temperature, pressure, number of deuterons per unit volume, radial distance to the container wall, speed of propagation of the flame, rates of energy release in the thermonuclear material* and in the uranium wall, etc. Each of these quantities depends on all the others. Putting their interrelationships into mathematical form leads to what are called coupled differential equations. The basic equations don't change as the computing power increases. What changes is how many simplifying assumptions one has to make, what factors one decides to ignore, by what time interval one jumps from one moment to the next, how finely one divides up space, and so on. The goal is to complete a meaningful calculation in a reasonable time, given the computer's limitations of memory and speed. (The human computer had a limitation only of speed, not "memory," since paper was plentiful.)

In my first month or so in Los Alamos (summer 1950—a year before Matterhorn got going), I, along with other members of Wheeler's small team, focused on studying and cal-

---

*Adding to the complexity is the fact that some of the energy released in one place is carried by neutrons to some other place. By contrast, in chemical combustion, all of the energy released in one place is deposited at that place.

culating a variety of aspects of the classical Super and other thermonuclear ideas such as the alarm clock. (We also found time to start modeling the "shots" scheduled for the next May in Enewetak—see below.) I was, as the saying goes, getting up to speed, and was helping Wheeler and Teller prepare for their major presentation to the AEC's General Advisory Committee that September. By some time in August, as I described in Chapter 10, we had put together a large report, "Thermonuclear Status Report Part I," officially authored by Teller and Wheeler, and, as I mentioned earlier, unofficially known as the "telephone book" because of its bulk. [2] With suitable scientific caution, the report said, "As of August 1950 it is still impossible to say whether or not any thermonuclear weapon is feasible or economically sensible." Despite that judgment, Teller and Wheeler were very far from advocating a slowdown or hiatus in thermonuclear research. Just the opposite. They attributed the uncertainty to an insufficiency of research. They said, in effect: With more scientific talent, especially in theoretical physics, and with more time and more calculating, there is a good chance that a thermonuclear weapon can be successfully designed.

A strength of the report was that it provided a good summary of all that was then known or hypothesized about thermonuclear weapons. (Recall that this was pre-radiation implosion.) From my personal perspective, pitching in to help prepare it was a great way to get rapidly into the subject that was to be the focus of my professional attention for the next two years.

In its report following its September meeting, [3] the GAC cautioned against letting the fusion program at Los Alamos drain too many resources from the fission program, but, as I noted in Chapter 10, also praised the thermonuclear efforts and recommended that the AEC provide more computing resources for those efforts (was it in the hope that more calculations would prove, once and for all, the impossibility of building the weapon—who knows?). Oppenheimer's reaction was to express his "frustrated gratitude" to Teller and Wheeler and their troops. [4]

By the time of the "telephone book" report, Los Alamos, guided by its Family Committee, had decided on a test of thermonuclear burning the next spring, in what would be the George shot of the Greenhouse test series. The unusual cylindrical design of George's fission-bomb trigger allowed for quicker flow of radiation from the fission bomb along its axis to the capsule of deuterium-tritium (DT), whose ignition was the goal. [5] The purpose of the design was only to get the energy promptly to the DT where it was needed, not specifically to compress the DT. As it turned out, the test was entirely successful, and even provided evidence that the radiation had caused some implosion of the DT capsule. The result of the George shot was very satisfying to all who had worked so hard to make it happen, yet it was not exactly cause for euphoria either, because by the time it took place, the idea of radiation implosion was in the air, and we were all optimistic that there was now a clear path to a workable weapon, even without the reinforcing evidence of the George shot.

What the George shot did do was provide what was very likely the first example on Earth of fusion triggered by heat. DT reactions, with their resulting 14-MeV neutrons, were commonplace in accelerator laboratories. This was the first time that those telltale neutrons were propelled from a reaction not as the result of an accelerator beam causing deuterons and tritons to fuse, but because of heat so intense that the D's and the T's acquired thermal energy sufficient to cause their fusion.

The Item shot, a couple of weeks after George in May 1951, also involved DT burning (and the concomitant release of high-energy neutrons), but, as I described in Chapter 10, in quite a different configuration. The thermonuclear fuel, instead of being off to one side of a fission bomb, rested in a central cavity within a spherical fission bomb. This was the "boosting" principle, in which the purpose of the DT burning is not so much to add fusion energy as to intensify fission energy. It, too, was entirely successful. [6]

In the months following the marathon effort in the summer of 1950 to prepare the report for the GAC, I spent time on calculations specific to George and Item, both of which were largely designed by the fall of 1950, and also pursued burning calculations on cylinders of deuterium and on other configurations such as deuterium bubbles in uranium or plutonium ("Swiss cheese"). By August 1950 Fermi and Ulam, using hand calculations (carried out largely by Miriam Planck and Josephine Elliott), had confirmed the pessimistic outlook of the earlier Ulam-Everett calculations for the classical Super. This dampened but certainly did not extinguish our enthusiasm for what we were about. Wheeler, like Teller, had an abiding faith that what was possible in principle could be—and would be—achieved in practice if we just kept scratching for new ideas and exploring a lot of them. Wheeler's young colleagues—John Toll, Burt Freeman, and I—took our cues from our elders (Wheeler was all of thirty-nine and Teller was forty-two). We went at our calculations and modeling and theorizing with a sense that we would triumph if we worked hard enough. It was not unlike grappling with a particularly challenging homework problem.

Preparing materials for the human computers was my introduction to programming. Instructing these young women was actually not so different from instructing a mechanical computer. One had to lay out the arithmetic steps in detail, not only commands such as "multiply A times B and subtract C," but also conditional commands such as "If the result of A times B is greater than D, go to step E; otherwise go to step F." So my transition to programming for the IBM CPC (card-programmed calculator) was straightforward, and by the midpoint in my year at Los Alamos, I was regularly using CPCs for numerous calculations. But I wasn't the only one. Others working on thermonuclear reactions, such as John Toll, as well as the team developing new fission weapons all came to rely more and more on the CPCs. Our "elders" counted on us "youngsters" to do the actual programming.

The CPC, first offered for sale by IBM in 1949, [7] was a modified accounting machine, about the size and weight of a commercial refrigerator. Instructions and data were fed into it with a stack of punched cards, each card having eighty columns and each column having space for up to a dozen small rectangular punches. A stack of, say, a hundred cards, would be placed in a hopper, then sucked into the machine at the rate of a little more than one per second. If there was a break of a few seconds in the otherwise regular click-click-click of the card reader, it meant that the machine was pausing to do something complicated like taking a square root. The cards emerged into a lower hopper from where they could be lifted back into the input hopper. At any given moment, there were only one or two cards from the input stack inside the machine. Still in the future—the very near future, in fact—were the so-called stored-program machines in which instructions residing within the machine directed the computer's operations without further individual intervention.

The results of the CPC's calculations could be punched onto new cards or printed onto fifteen-inch-wide continuous-feed form paper. The printer, already mechanically sophisticated because it was part of the previously developed accounting machine, could provide a row of up to 132 digits in a single stroke. Between the input and the output came the arithmetic. That was controlled by plug boards a little less than one square foot in size that could be individually wired to meet the needs of a particular kind of calculation. For instance, suitable wiring could allow a single command on a punched card to order a square root, leading to the kind of hiccup mentioned above.

An IBM card-programmed calculator (CPC) in use in a Seattle municipal office in 1954. The hopper for a stack of cards is just to the left of the printer. Well after CPC's ceased to be of interest for scientific calculations, they remained useful in other applications. *Courtesy of Seattle Municipal Archives via commons.wikimedia.org.*

The CPC was slow, ponderous... and reliable. The ENIAC, a nightmare to program and already obsolete, had served its evolutionary purpose. Princeton's MANIAC, revving at the starting gate and ready to launch a revolution, was to the CPC what a prototype Aston Martin is to a Model T. It was fast, sleek... and unreliable. The new breed of stored-program machines, offspring of the MANIAC, would soon take over, but for a year or two, the CPC did its job, churning out the calculations that kept the nuclear weapons programs on track. (And not just the weapons programs. Of the more than 500 CPCs delivered by IBM to government and industrial concerns, many, like antiquated Model T's, kept on serving the needs of their owners for many years.[8])

There were, if I recall correctly, only three or four CPCs in Los Alamos during my year there. By the spring of 1951 this was not enough to meet demand, even with late-night use of them.

So, shortly before my return to Princeton, I was dispatched to Sandia Labs in Albuquerque to carry out calculations there, where CPC usage was not yet saturated. At least not saturated around the clock—time in the middle of the night was made available to me. For a place to sleep during the day I was assigned a room in the BOQ (bachelor officer quarters), which gave me a certain wry pleasure since as an enlisted man in the Navy half a dozen years earlier I had perceived officers as light-years distant.

These Sandia calculations, like the ones just before at Los Alamos, were on the propagation of a thermonuclear flame in a cylinder of deuterium that had undergone radiation implosion (using a lot of hypothetical numbers, since the degree of compression and even the dimensions of the cylinder were unknown). The results began to look encouraging.

No sooner was I installed in Princeton than the Los Alamos authorities arranged for me to continue these calculations on a CPC located in an IBM building on the East Side of Manhattan (not the principal IBM building in midtown). I was—of course—assigned an extended graveyard shift, 8:00 p.m. to 8:00 a.m. if I remember correctly. Typically, I took the train from Princeton Junction to New York in the evening and returned by train in the morning, checking in with my colleagues before seeking some rest. To this day I have fond memories of the Hamburger Heaven across the street from the IBM building. It stayed open all night. And good memories of John Sheldon, the aspiring executive who ran this small branch of IBM. He constantly fussed because IBM culture dictated that he wear a hat on his way to and from work. On his staff was a blind young man who expertly wired plug boards.

Chapter 14
The Garwin Design

An important meeting of an advisory panel convened by the Atomic Energy Commission was scheduled for Saturday and Sunday, June 16 and 17, 1951, and was to be held at the Institute for Advanced Study in Princeton, fortuitously soon after Matterhorn got going. It was a truly "blue-ribbon" panel, attended by all five Atomic Energy Commissioners;* Robert Oppenheimer and four other members of the General Advisory Committee; a Los Alamos contingent that included Norris Bradbury, Carson Mark, Lothar Nordheim, and Edward Teller; John Wheeler from Matterhorn; and what I call the "big-three" consultants: Hans Bethe, Enrico Fermi, and John von Neumann—in total about twenty people. [2] Progress toward a thermonuclear weapon was to be a principal agenda item.

I spent that Saturday night in New York nursing a CPC, as I had spent every night that week. I tweaked the input parameters and assumptions this way and that way and could not escape the conclusion that the compressed deuterium would probably burn. On Sunday morning I took an earlier-than-usual train to Princeton Junction and carried the computer output with me to the Matterhorn building. After the guard let me in I translated the numerical results I was carrying into a couple of graphs inked with marker pens on oversized paper, perhaps as large as two feet by three feet. By now it was after 9:00 a.m. I rolled up the graphs, got back in my car, and drove to the Institute for Advanced Study, where the advisory panel

---

*Sumner Pike, an original Commissioner appointed in 1946; Gordon Dean, Chair, and Harry Smyth, both appointed in 1949; and T. Keith Glennan and Thomas Murray, both appointed in 1950. [1]

meeting was taking place in a first-floor conference room. I walked across the grass to the low window of the meeting room and saw John Wheeler standing, as if about to deliver some remarks. Either by gesture or by tapping on the window, I caught his attention. He walked over to the window and opened it. I handed him the graphs, saying "Looks good." He went back to the front of the room and taped the graphs to a chalkboard. I went in search of breakfast and some rest.

My graphs reinforced whatever message Wheeler and Teller were delivering to the assembled group. This was, apparently, the meeting at which the GAC—and, in particular, its chairman, Robert Oppenheimer—switched from doubt and questioning to a conviction that the "new" Super would very likely work. Oppenheimer famously called the Teller-Ulam approach to a thermonuclear weapon "technically sweet," a perspective that may date from this 1951 meeting. And, who knows, one that my graphs may have helped to solidify.

(I should explain the context of Oppenheimer's "technically sweet" comment. At the 1954 hearing that resulted in the loss of his security clearance, he was being questioned about his 1949 opposition to going forward with a program to develop an H bomb. To his interrogator he said, in effect, that he felt free to oppose it on moral grounds at that time because, on technical grounds, it seemed doubtful that it could be done. It was his judgment, he said, that "when you see something that is technically sweet, you go ahead and do it and you argue about what to do about it only after you have had your technical success. That is the way it was with the atomic bomb." [3] By inference, he was saying, had we felt confident in 1949 that the H bomb could be built, we would have been less likely to oppose it, and instead would have waited until its completion before deciding what to do about it. His actual words were "I cannot very well imagine if we had known in late 1949 what we got to know by early 1951 that the tone of our report would have been the same." Later in his testimony, and still in response to challenges about his 1949 opposition, he said, "The program we had in 1949 was a tortured thing that you could well argue did not make a great deal of technical sense. It was therefore possible to argue also that you did not want it even if you could have it. The program in 1951 was technically so sweet that you could not argue about that." [4])

Richard Garwin in the early 1950s. *Courtesy of Richard Garwin.*

In 1951 Dick Garwin came for his second summer to Los Alamos. He was then twenty-three and two years past his Ph.D.* Edward Teller, having interacted with Garwin at the University of Chicago, knew him to be an extraordinarily gifted experimental physicist as well as a very talented theorist. He knew, too, that Fermi had called Garwin the best graduate student he ever had. [5] So when Garwin came to Teller shortly after arriving in Los Alamos that summer (probably in June 1951) asking him "what was new," [6] Teller was ready to pounce. He referred Garwin to the Teller-Ulam report of that March and then asked him to "devise an experiment that would be

---

*In the previous summer, when Garwin and I overlapped in Los Alamos, he had paid me a compliment by being surprised that I was still a student.

absolutely persuasive that this would really work." Garwin set about doing exactly that and in a report dated July 25, 1951, titled "Some Preliminary Indications of the Shape and Construction of a Sausage, Based on Ideas Prevailing in July 1951," [7] he laid out a design with full specifics of size, shape, and composition, for what would be the Mike shot fired the next year.

Garwin then presented his design to the Lab committee overseeing thermonuclear developments, a committee chaired that summer by Hans Bethe, visiting from Cornell. (This may have been an interim committee functioning between the Family Committee and the Theoretical Megaton Group).* Bethe, an eminence in the physics world whose words were seldom questioned, said that he thought the walls of the outer casing in Garwin's design were not thick enough. Garwin, to the astonishment of some who were present, said, in effect, "You are wrong, Hans, and here's why." Bethe admitted the soundness of Garwin's argument, and the design was approved. [9] From then on, we at Matterhorn, as well as the fission-bomb designers at Los Alamos and a slew of other specialists, knew what we had to work with.

Regarding Teller's faith in the young, I must relate one example in which I played a part. Some time in early 1951 Teller came to me holding what looked like a thick report in his hand. "Ken," he said, "I am getting redder [with embarrassment] by the microsecond. I have this draft dissertation from my student in Chicago, and I haven't had time to read it. Would you please read it and tell me what you think." I read it with care. It was a thorough discussion of a proposed experiment to search for the hypothetical magnetic monopole. (A monopole is a particle analogous to an electron that carries magnetic instead of electric charge. By chance, as I related in Chapter 11, I had done some theoretical work on the behavior of this

---

*The last meeting of the Family Committee for which minutes are preserved took place in June 1951. The first official meeting of the Theoretical Megaton Group took place in late September or early October 1951. [8]

supposed entity. To date there have been many searches for it and no sightings.) I found the dissertation to be admirably clear, and free of any errors so far as I could tell. I reported this to Teller and he then notified the student of his approval (no doubt after examining the dissertation himself).

What is the Garwin design? What is a "sausage"? The second question first. The outer steel capsule containing the entire device had a length about three times its diameter, so it roughly resembled the proportions of a fat sausage—at some twenty feet in length and nearly seven feet across, quite a large sausage. See its picture on page 180, with a man and a Jeep to set the scale. (I once heard someone say that the name was adopted in part because cylinders of thermonuclear fuel could be joined end to end like sausage links. That etymology is questionable, since the term "sausage" was adopted immediately for a single cylinder and appeared in the title of the April 1951 report written by Freddie de Hoffmann and signed by Teller. [10] Still, one must note that in both de Hoffmann's home city of Vienna and Teller's home city of Budapest, linked sausages offered by street vendors were common sights.)

As to other details, they are, technically, still secret. However, among the many unclassified published accounts, there is a broad agreement about the design of Mike. I report that consensus view here. At one end of the long cylindrical steel container is the fission bomb that will provide the radiation and get the thermonuclear process started. The fission bomb is the "match" that will light the rest. Working in from the outside of the steel cylinder, there is a layer of mostly low-density material such as polyethylene, which provides an easy channel for the radiation. (In his first design, Garwin proposed liquid hydrogen in this space, because he knew better how to calculate its behavior when flooded with radiation. [11]) Then comes a cylinder of ordinary, non-enriched uranium—reportedly five tons of it. Within that is a huge stainless-steel "thermos bottle" (a dewar) containing the liquid deuterium that is the thermonuclear fuel. That "bottle" includes evacuated layers to inhibit heat flow from the deuterium, which is at a temperature of just 24 degrees above absolute zero (about -249 degrees Celsius). There is also cryogenic "plumbing" to maintain that low temperature.* And finally, running along the axis is a slender cylinder of plutonium 239 (the highly fissionable isotope)—subcritical, of course until compressed. Within it, according to some reports, is yet a final detail—a pencil-thin space at the center containing a very small quantity of a deuterium-tritium mixture, just enough to "boost" the plutonium fission by providing additional neutrons from the DD and DT reactions. This axial plutonium is called the sparkplug.[†] [12]

What then happens when the fission bomb "match" explodes? Its radiation runs out ahead of its expanding material and almost instantaneously vaporizes the polyethylene, creating a very hot plasma. The pressure of this plasma adds to the pressure of the radiation, pushing outward on the outer steel cylinder and at the same time inward on the uranium cylinder, thereby keeping the channel open long enough to let more radiation stream in. At the same time, the outer layers of the inner cylinder "ablate" (boil off), creating even more inward pressure. Before long—within microseconds—an inwardly imploding shock wave has compressed and heated the cylindrical container of deuterium and also, through compression, caused the plutonium sparkplug to go critical. The sparkplug, which is really another fission bomb, helps to ignite and then enhance the deuterium burning—hence the name

---

*When working at Los Alamos, I turned down every invitation to see an actual bomb or mockup of a bomb. I didn't want the wires and bolts and plates to intrude upon my visualization of perfect cylinders and perfect spheres. I thought that seeing the real thing might make me less effective as a designer.

---

†Both of the two cited references in note 12 say that the central axial cavity in the spark plug contained tritium gas. Tritium alone would not have been helpful. It must have been a DT mixture as in boosted weapons.

"sparkplug." Finally, energetic neutrons emitted in the thermonuclear burning (especially those of 14 MeV from the DT reaction) cause fission in the ordinary uranium that surrounds the
deuterium as well as more fission in the sparkplug. Altogether
quite a maelstrom. So this "H bomb" is really one part fusion
and three parts fission. Its total energy release (its "yield") is
estimated, in fact, to have come about three-quarters from
fission and one-quarter from fusion. [13]

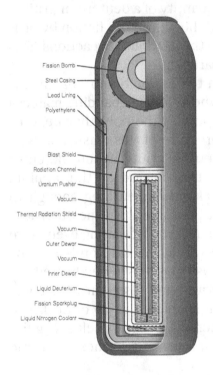

Fission Bomb
Steel Casing
Lead Lining
Polyethylene

Blast Shield
Radiation Channel
Uranium Pusher
Vacuum
Thermal Radiation Shield
Vacuum
Outer Dewar
Vacuum
Inner Dewar
Liquid Deuterium
Fission Sparkplug
Liquid Nitrogen Coolant

An imagined structure for Mike, based
on what appears in Richard Rhodes's
book *Dark Sun* (p.492) and on the
Web at http://www.sonicbomb.com/
content/atomic/carc/us/ivy/limg/
mikedevice.jpg

Everything that I have just described happens before
the entire assembly blows itself apart. Fortunately for the
bomb-builder, the big chunks of steel (nearly fifty tons of it)
and uranium (perhaps five tons) are sufficiently "sluggish" to
allow much of the fission and fusion to run their courses before everything is scattered to the winds. It doesn't take many
microseconds.

So, beginning in the fall of 1951, we at Matterhorn and our colleagues in Los Alamos had a clear view of the device whose behavior we wanted to calculate. We also had about six months in which to suggest design changes, if any, since Mike's design would need to be frozen in the spring of 1952. One change that we all knew we wanted, eventually, was to use lithium-6 deuteride rather than deuterium as the thermonuclear fuel. But this was for the more distant future. Calculating the behavior of deuterium was easier and surer. The less uncertainty the better. For a test, all of the complexity of handling liquid deuterium and keeping it super-cold was worth the trouble.

In 1951 and 1952, we at Matterhorn and at least one Los Alamos scientist, Marshall Rosenbluth (like Garwin, another young star from Chicago), did carry out some calculations of thermonuclear burning for compressed "sausages" of lithium-6 deuteride, but this was with an eye to the future beyond Mike. Our principal focus remained on deuterium.

At the same time, it wasn't fully ruled out that deuterium might be used in an actual deliverable weapon. Here is Dick Garwin's laconic remark about what he did in the remainder of the summer of 1951 after designing Mike. [14] "After July 25th, I had some time at Los Alamos, and so I designed a deliverable, cryogenic version of Mike that would lie down rather than stand up." (Mike was vertical.) No easy task. To meet Air Force requirements for dealing with air turbulence or hard landings, the bomb had to withstand forces up to eight "g's" (eight times its weight). It had to be no heavier than a B-36 bomber could carry, about 35 tons. (It ended up closer to 20 tons.) And its load of deuterium had to remain frigid and liquid for many hours without mechanical refrigeration. On the ground, until takeoff time, it could be kept cold with ground-based cryogenic equipment. In the air, until its release or until the bomber returned to base, it was probably surrounded by a larger dewar filled with liquid hydrogen (ordinary hydrogen, not deuterium)—a thermos within a thermos, so to speak. Garwin had to design it so that during a long flight the amount of deuterium allowed to escape as gas would be a small fraction of the total and the pressure within the "sausage" would be tolerable. Garwin was up to the task. His design included even innovative bolts that minimized heat flow out of the stainless-steel dewar.

About half a dozen of these "recumbent Mikes" were actually built and were ready for deployment by January 1954. [15] They were called Emergency Capability Weapons, officially designated TX-16's, and unofficially nicknamed Jugheads. But they had hardly come into existence before they were rendered unnecessary and obsolete by the successful test on March 1, 1954 of a solid-fueled H bomb, one containing, instead of deuterium, lithium deuteride enriched in the isotope lithium-6. This Castle Bravo device, called Shrimp (who knows why?), released fifteen megatons of energy, a thousand times that of the Hiroshima bomb and, in fact, more than twice what was predicted. Reportedly, the Los Alamos designers failed to take account of the significant contribution of the isotope lithium-7 to the thermonuclear reaction. [16]

Shrimp's unexpectedly large yield plus an unanticipated wind shift produced the worst ever fallout damage from a U.S. nuclear test. Residents of Rongelap and Rongerik atolls in the Marshall Islands were evacuated (two days later, after already suffering significant radiation harm), and the crew of the Japanese fishing boat "Lucky Dragon No. 5," who very unluckily found themselves in the downwind fallout path, experienced serious radiation sickness. One man died, and the world took notice. [17]

In my two years as an apprentice weapons designer, I worked hard at my job (and derived a good deal of satisfaction from it). Yet at the same time I managed quite a

bit of relaxation. Apart from square dancing and folk dancing, much of the relaxation involved wheels. There were the
weekend outings from Los Alamos with other young people,
the drive with Mici Teller, Janette Wheeler, and John Toll to
Los Angeles, and the motorcycle ride across the country.
Over Thanksgiving 1950, I decided to get away from the lab
for a few days by driving my Carryall to Tucson and enjoy-
ing the southern New Mexico and Arizona scenery along the
way. The only activity I had planned for Tucson was reading
books. (I knew no one there.) By chance I stopped for gas in
Socorro, New Mexico, a town to which I would move twenty
five years later. At the time of my Tucson trip, a gas station
just south of town stood between the road and the air strip.
It pumped gas for cars on one side and gas for planes on the
other side. By the time I lived in Socorro, I had become a
pilot (and drove a Volkswagen). At the airstrip there was no
longer service for cars, but I took pleasure in gassing up the
planes I flew next to where I had gassed up my car several
decades earlier.

On one of my train trips from Princeton to Los Alamos
in the winter of 1951–52—probably in December 1951—I re-
captured the Carryall that I had left behind and drove it back
to Princeton. Since it had no heater, I opted for a southern
route—across the seemingly endless miles of Texas, then
through Mississippi, Alabama, and Georgia, and on up through
North Carolina and Virginia to Washington, DC, and on to
Princeton. As it happened, the weather that week was odd.
The Deep South was in a deep freeze. It was warmer in Illinois
and Indiana than in Mississippi and Alabama. I drove most of
the trip wearing winter gloves and a heavy coat.

Not long after reaching Princeton, I skidded on the ice
on Prospect Street and slid into the rear of a large truck. The
result was a baseball-sized hole on the right side of the Carryall that thereafter funneled rain and snow onto the right knee
of any passenger on the far right of the front seat. Some of

my passengers took this brush with nature stoically. Some slid across the seat to the left. A few asked politely to get out.

Several months after the encounter with the truck, I was driving to Washington, D.C. on a rainy night with John Toll on board. On a stretch of four-lane highway I lost control of the car and it spun around through 180 degrees. Toll, who had been resting on the mattress in the back of the Carryall, was jarred awake. He said later, "I sat up and looked out and wondered why we were going backward." I got the car pointed in the right direction and we proceeded without further incident to Washington—where Toll's family lived and where I had a date with a computer. The gods are kind to the young.

The last I heard of the Carryall it was in Winnipeg. In the spring of 1953 I sold it (for $75) to my fellow graduate student and then-roommate Ken Standing, a Canadian who went on to a distinguished career as an experimental physicist. Ken drove it to his new job in Winnipeg where, I hope, he bought a heater for it.

Chapter 15
Climbing Matterhorn

For some reason, the number $20,000 has stuck in my head over the years. It is the amount I heard that the AEC was paying the National Bureau of Standards per month for the graveyard shift on the SEAC. That was about 50 times my salary at the time, so it was memorable.

The night shift was my lot in life, and I didn't mind. For some months in the spring and summer of 1952, I rented a basement room just off Connecticut Avenue, not far from the Bureau of Standards, wrestled with the SEAC at night, and slept part of the day. (I had to keep a notepad by the phone, which was next to my head. Had I not scribbled a note after every call that woke me, I would have gone right back to sleep without remembering a thing about the call when I got up in the afternoon—even though, to the caller, I probably sounded wide awake.) I hired a young man with no scientific credentials as an assistant. His job was to operate the computer and keep track of its input paper tapes and its output pages printed by a teletype machine. I trained him to convert the computer's hexadecimal (base-16) output into decimal numbers so that we could summarize our results in graphs and tables. He had no clearance and at best (I hope) a very hazy idea of what we were doing.

The purpose of the SEAC calculations was simple: Follow the thermonuclear burning and the related fission for "steady-state" flame propagation in the Mike device and calculate Mike's total energy release (its "yield"). A few "runs" had to be duplicated by hand to make sure that the computer was doing what we thought it was doing. (The time is now long past when checking a computer's work by hand is possible.) Also most runs had to be repeated to be sure that two calculations with identical input gave identical answers. And we had to do run after run after run with altered input data, altered assumptions about the physical processes, and altered calculational approximations.

Even though the SEAC was, at the time, probably the best computer in the world, its limitations still dictated a calculational approach that was highly simplified. We assumed that the fission trigger had done its job, creating a bath of super-hot radiation; that this radiation had compressed the deuterium (and its central "sparkplug") to a targeted compression; that thermonuclear burning had been initiated; and that a flame was now spreading through the material. These starting "assumptions" were really the end points of other calculations—calculations that were intended to assess the early stages of the explosion. Some of these were carried out in Los Alamos, some at Matterhorn. My SEAC calculations were designed to find out if the flame, once started, would continue to propagate through the material (in a "steady state") or would fizzle, and, if it didn't fizzle, what total energy would be released.

The calculations that I labored over night after night and the equations and input assumptions that I tinkered with on many a day, when added to all of the other theoretical and calculational efforts being conducted by a very small army of physicists and mathematicians (no more than a few dozen of them), convinced us finally that Mike would be a success in the multi-megaton range. I save our final predicted number and its comparison with the actual yield for the next chapter.

As I mentioned at the end of Chapter 11, when I got back to Princeton from Los Alamos in 1951, I did what other young people tend to do with accumulated money: I spent it. The two-seat British Singer Roadster that I bought had the look and feel, if not exactly the power, of a genuine sports car. It "cornered" very nicely, and it had a feature especially suitable to Washington's hot, humid summer. Not only did the canvas top come down, but also the windshield folded forward. I could speed about the streets of Washington with the wind in my face. One girlfriend at the time liked to ride in the Singer crying out "faster, faster." We both survived and she married a doctor who drove a sedan.

The Singer was handy for the frequent trips that I had to make back to Princeton. (The Carryall was relegated to the status of "second car.") John Wheeler was the unquestioned leader of the physics we were exploring, and I needed to check in with him often. Occasionally he came to Washington, but mostly I went to Princeton. John Toll was also heavily involved. He and I assisted Wheeler in developing the equations that we wanted to use. Toll also worked with me in writing programs for the SEAC. Wheeler took no direct role in the programming but was keenly interested in whatever approximations we had to make.

The SEAC was closely modeled after John von Neumann's MANIAC. (Indeed every computer since owes some features of its architecture and operation to that progenitor machine. [1]) The main difference was that the SEAC had two parallel memory banks, whereas the MANIAC had only one. The MANIAC's memory consisted of what were called Williams tubes*—forty of them. These were cathode ray tubes in which an electron beam scattered charge away from spots on the screen (a 32 × 32 array of 1024 spots) and then could determine by a complex

---

*The Williams tube was invented by the British electronics engineer Frederic C. Williams. [2] The invention is sometimes attributed to both Williams and Tom Kilburn and accordingly is known also as the Williams-Kilburn tube.

dance that included a redirected electron beam and an adjacent metal plate whether or not that spot's charge had been scattered—thus whether it represented a zero or a one. (And every time a spot was read, it had to be instantly rewritten.) This memory was fast but it was balky. The electrons didn't always do what they were supposed to do.

A little arithmetic will tell you that the MANIAC'S memory amounted to about 40,000 bits, or about 5,000 bytes. Now, for around ten dollars, you can purchase a "thumb drive" with a memory capacity greater than that of a million MANIACs.

The SEAC also had a bank of Williams tubes (with about half the memory capacity of the MANIAC) and in addition a set of what are called mercury delay lines. Each delay line looks a bit like a fluorescent tube, two feet long and a little more than an inch in diameter. It doesn't emit light. Instead it sends little acoustic pulses from one end to the other and then feeds these pulses electrically back to the starting point. So zeros and ones are constantly running along the tubes: 360 of these bits chasing each other down each tube. As they emerge from the ends of the tubes, they can be "read" and copied into the computational part of the computer. For better or worse, the two kinds of memory could not be used at the same time. There was a simple switch on the SEAC: Williams tubes in one position, mercury delay lines in the other position. We usually had the switch in the delay-line position, to assure more reliable, albeit slower, operation. Every once in a while we would throw the switch to the Williams tube position and hope for the best.

Every reader of this book is likely to use a computer—desktop, laptop, tablet, or smart phone. I will therefore take a little more space to describe features of the SEAC, to contrast it with modern computers.[3] (I gained a true affection for the SEAC, which was, after all, my nighttime companion for the many weeks of a long, hot summer.) It had 512 "words" of memory. Each word, 45 bits in length, could accommodate either one number of up to 13 digits or one command. In practice, the large majority of the words were given over to commands. It was normally sufficient to store no more than a few dozen numbers. The SEAC—unlike nearly all computers since—was a "four-address" machine. This means that a single command line might contain instructions such as the following: Take a number from address A and a number from address B, combine them in some way or compare them, send the result to address C, and then go to address D to get your next command. (Forty-five bits is enough for all of this.) Needless to say, programming was in "machine language." I would sit down with a few sheets of paper containing 512 numbered lines, and write the commands and the designated numerical storage locations on the numbered lines.

Some of the SEAC's wiring. Pulsing its electrons was a clock ticking at 1 MHz, less than one-thousandth the speed of the clock in a modern laptop. *Courtesy of National Institute of Standards and Technology Digital Collections, Gaithersburg, MD 20899.*

The commands, once written on paper, were then transferred to holes punched in a paper tape and fed into the machine via a teletype machine. That teletype machine, in turn, served to print out results of the computations—in hexadecimal, as I noted earlier: A number might look like 94B7C35AF0.

I wasn't the only H-bomb theorist using the SEAC that summer of 1952. Marshall Rosenbluth, one of the brightest young physicists at Los Alamos, came for a little while to do runs comparing the solid thermonuclear fuel lithium deuteride and the liquid fuel deuterium.* He, assisted on at least one occasion by Dick Garwin, was looking ahead, beyond Mike. As for me, my sole focus that summer was on the upcoming Mike test scheduled for the fall. Only after Mike was "put to bed" did I branch out, considering later assemblies, such as the Castle Bravo shot (with solid fuel) planned for the spring of 1954. But soon after the successful Mike shot, which produced in me that odd combination of euphoria and dread that other nuclear weaponeers have no doubt also experienced, I holed up in my small office in Palmer Lab and got to work on my doctoral dissertation research.

Matterhorn functioned for less than two years. In the project's final report, [6] prepared in August 1953, Wheeler gives these dates:

- Late December 1950. The project was "conceived" "to help Los Alamos overcome [a] shortage of theoretical

---

*Rosenbluth mimicked Garwin in precocity. He finished high school at 15, earned his Harvard bachelor's degree at 19, and finished his Chicago Ph.D. at 22. [4] Teller was successful in bringing him to Los Alamos in 1950, where he remained for six years. Like Stan Ulam, he found the mixture of applied and pure research to his liking, and continued that balance at General Atomics before joining academia. His later honors included the National Medal of Science in 1997. He died in 2003. [5]

manpower." (This probably means the time when the Los Alamos authorities gave it their blessing; it was conceived in Wheeler's mind earlier).

- Late February 1951. Los Alamos and Princeton University informally agreed to go forward with the project.
- May 21, 1951. Work began. (This is the official start date. It may have been a week or two later when Wheeler and Toll and I set up shop in the just-acquired metal shack, slept in the Forrestal boiler room, and began recruiting and hiring support staff).
- June 28, 1951. Princeton University and the University of California signed a contract formally establishing the project.
- Early March 1953. The Project was terminated. (Wheeler and a few others put in time thereafter to wrap up the project and prepare its final report.)

I have already commented, in Chapter 11, on the stunning speed with which Matterhorn went from an idea to reality, a speed made possible by fear of the Soviet Union and the related near-hysteria of anticommunism at that time. Equally remarkable is the speed with which Matterhorn, once established, became a significant functioning organization making important contributions to thermonuclear designs. From boiler-room bedroom to full partnership with Los Alamos no more than two or three months elapsed.

. Why was Matterhorn able to "hit the ground running"? In part just because of that partnership with Los Alamos. It was as if we were another group in T Division—albeit larger and more independent than the other groups. And in large part because of John Wheeler's drive and leadership. He was never faint of heart. When he could not induce senior physicists to join the project, he rounded up a group of very bright young physicists and personally inspired their work. When he found that the MANIAC at Princeton's Institute for Advanced Study—around the corner, so to speak—was not ready for serious, reliable work, he quickly arranged for us to use an IBM CPC in New York, the SEAC in Washington, and a UNIVAC in Philadelphia. And sprung loose the money to make that possible. Arrangements for the UNIVAC included postponing its delivery to the government agency that had ordered it so it could first serve the higher-priority interests of the Atomic Energy Commission. (According to my memory, it was the Weather Bureau that had to wait, but it may have been the Census Bureau.)

As I mentioned on page 128 only Louis Henyey from Berkeley responded favorably to the invitation to join Matterhorn that Wheeler sent out to a great many senior people in the spring of 1951. (In his 1998 autobiography, Wheeler puts the number of people contacted at 120. [7] In his 1953 final report on Project Matterhorn's work, he says that he wrote to "all known suitable U. S. theoretical physicists and applied mathematicians—a total of more than 140 people." [8]) As I recall from the time, very few of those contacted declined because they wanted nothing to do with weapons work. Instead, the reasons included an unwillingness to leave academic work so soon again after the end of World War II; an established position doing defense work elsewhere; and an insufficient proposed salary at Matterhorn. [9]

Among those contacted by Wheeler was, quite naturally, Richard Feynman, his most brilliant former student and his colleague on important work in the 1940s. Wheeler's letter of March 29, 1951 to Feynman begins: "Dear Dick: I know you plan to spend next year in Brazil. I hope world conditions will permit. They may not." He continues with the same two-page "full-court press" used in his many other letters. Here is Feynman's response of April 5 in full:

> Dear John:
> As you know, I was planning to spend my sabbatical leave in Brazil. I am uncomfortably aware of the very large chance that I will be unable to go. Until that situation be-

comes definite, however, I do not wish to make any commitment for work next year.
    Best wishes, Richard P. Feynman. [10]

Brazil enabled Feynman to avoid saying whether he would or would not want to join Matterhorn, whether he did or did not approve of Wheeler's project.

Wheeler's final Matterhorn report [6] consisted of six parts:

- History of the Project
- Documentation of its work
- Compression
- Burning
- Weapons of the third kind
- Unknown subject

With his characteristic flair, Wheeler arranged the six parts like stories in a daily newspaper, all starting on page 1 and continuing onto various back pages (of which there were 142). The opening "newspaper" spread appears in the illustration on the next page. It shows Wheeler's imaginative presentation, complete with a quote from Goethe (with one word misspelled), and shows also the result of the redactor's hand more than sixty years later. On the "inside" pages, only 20 percent of the material survives, and that is largely descriptive material, not technical material. Our Matterhorn work remains secret.

The authorship of PM-B-37 is ascribed to John Wheeler plus four of his young colleagues, including me.* To use Niels

---

*The other three are Edward Frieman, who became a professor of astrophysics at Princeton and later director of the Scripps Institute (he died in 2013); John McIntosh, who became a professor of physics at Wesleyan University; and H. Pierre Noyes, later a professor at the Stanford Linear Accelerator Center. We were all in our twenties when we worked at Matterhorn.

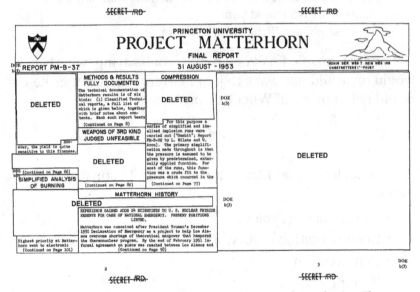

Cover page of Project Matterhorn's final report, PM-B-37, after redaction. The quote from Goethe's Faust in the upper right (with one misspelled word) translates as "Whither the path? There is no path where no one has gone before." *Secured from the U.S. Department of Energy through a Freedom-of-Information request. (Thanks to Fletcher Whitworth for assisting in the acquisition.)*

Bohr's terminology (see page 28), I feel innocent of the contribution. I must assume that Wheeler incorporated into the report some materials that I had written earlier.

One part of the report (designated "Unknown subject" above) is evidently so sensitive that the title, not just the content, is deleted. About "Weapons of the third kind," to which nineteen pages of the report are devoted, no clue remains as to what they were, and I cannot provide enlightenment. The reader learns only that, based on work by Pierre Noyes, they are judged to be unfeasible.

In the report, the subject of "Compression" is itself compressed to just five pages. This is because so much of the work on compressing the thermonuclear fuel was done in Los Alamos. Our focus at Matterhorn was burning (occupying 73 pages in the report). As I described earlier, my efforts on the SEAC dealt with the so-called steady-state burning in the late stages of the explosion. These calculations gained the code name Swordtail (I no longer remember the reason). At the same time, Larry Wilets was pursuing calculations on Philadelphia's UNIVAC concerned with earlier stages of thermonuclear burning. His calculations, code-named Chief (after the Santa Fe Railroad train, if I remember correctly) sought to answer the questions: Does the thermonuclear fuel ignite? Does a flame start to propagate? Chief could tell whether an explosion was likely. Only Swordtail could predict how much energy would be released (the "yield").

There were some 25 scientists and mathematicians (all male) who contributed for times short and long to Matterhorn. In his final report, Wheeler put the average number at any one time as about thirteen (see, for instance, the 1952 picture on page 141). Exclusive of the support staff, the project's backbone—those who worked for fifteen months or more—consisted of just eight physicists: Wheeler, Toll, and I, plus Walter Aron, David Carter, Pierre Noyes, Ralph Pennington, and Larry Wilets. Like soldiers in a platoon, we formed a close-knit group, and worked cooperatively toward a common goal, with little if any thought of competition or personal eminence. And, like soldiers, once our service ended, we mustered out and went on to other careers. As it turned out, those careers were almost all academic. Only a few of the twenty-five scientific staff members did later weapons-related work (mainly at the new Livermore Lab in California). Nevertheless, Wheeler dubbed us the "U.S. Nuclear Physics Reserve," able, because of our experience and clearability, to contribute further in case of national emergency. [11]

Matterhorn was, above all, focused. Focused on thermonuclear burning. Apart from one small but encouraging test at Greenhouse George in May 1951, thermonuclear burning existed on Earth only in the minds of theoretical physicists and in their notebooks and computer codes. Its big test was scheduled for Ivy Mike late in 1952. To be sure, we theorized about other configurations and about radiation flow and compression, but mostly it was all burning all the time.

The author at Project Matterhorn in 1952, preferring then, as now, to use a fountain pen.

One topic we left entirely to our colleagues at Los Alamos. That was the design of Mike's fission-bomb trigger, the device that would provide the flood of radiation leading to the implosion of the cylinder of deuterium. According to various reports, including that of the independent analyst Carey Sublette,[12] the choice for this fission bomb was the TX-5,* a spherical implosion bomb similar to, but smaller than, the Nagasaki bomb (smaller in size, not necessarily in yield). It had been used successfully in the earlier Greenhouse test series (and, according to public reports, became the basis for

---

*Also called TX-V, according to Richard Rhodes.[13]

deployed fission weapons for many years thereafter). [14] The main requirement for the fission trigger was that it not surprise. It had to predictably do what was expected of it. There were enough uncertainties in the rest of the Mike device.

Yet, surprisingly, the design *did* surprise Marshall Rosenbluth. In a 2003 interview, [15] he says that because he had been focusing on the thermonuclear part of Mike, he had not studied the fission primary. But this bright and energetic twenty-four-year-old wanted to understand all facets of the design.

> And so once before the test I went down and talked to the primary design group [at the Los Alamos Lab] and I found to my horror that in order to make sure it worked they had packed in so much plutonium it was sure to pre-detonate. So in fact a separate primary was flown out to the Marshall Islands a week or two before the shot.*

Rosenbluth's well-founded fear was that the TX-5 core, champing at the bit to go critical, would do just that before it reached full compression and would then explode anemically (by nuclear-weapon standards), bathing the sausage in much less radiation than was needed. Harold Agnew, a Los Alamos scientist who was later to become the lab's director, went as far as to say, "Marshall may have saved the Mike shot." [13] (To embellish this assertion, Agnew tells a story about Rosenbluth eating too much shrimp at a sumptuous dinner on Enewetak, not being able to sleep, and worrying about the fission core in his insomnia. A nice story, but Rosenbluth was not at Enewetak. [15])

Even after the design of Mike was frozen in mid-1952, we at Matterhorn could continue to refine our calculations of its performance and its yield. At the same time, the small army of experimental physicists and engineers who intended to take full advantage of the microseconds available to them during

---

*According to Rhodes's even more dramatic account, the new core arrive at Enewetak "a few days before the target date." [13]

the blast could continue to work away, revising the deployment of their wires and mirrors and electronic counters.

When October 1952 arrived, John Wheeler left for the Pacific, where he would observe the Mike test from a ship some twenty-five miles from the blast. He promised to report back to us. Edward Teller stayed in Berkeley but had his own plan for "observing" the shot.

Chapter 16

More than a Boy

There's another number that got stuck in my head—7 mega-tons. The sequence of events in Mike was quick, extremely quick, but it *was* a sequence, one thing after another. First the explosion of the fission trigger, then the flow of radiation, then the implosion of the "sausage," and finally the thermonuclear burning and the secondary fission. The last part of this sequence is what I calculated for many weeks of overnights on the SEAC. From all of this flowed a single number, 7 megatons, our predicted energy release, or "yield," of Mike. As it turned out, Mike flexed its muscles and did better than that.

Enewetak, like other atolls in the Pacific, is a chain of islands surrounding a large lagoon.[1] Located in the Marshall Islands about 2,700 miles west-southwest of Hawaii, Enewetak consists of forty islands (formerly forty-one) lined up in an oval some fifty miles around. Mike was only one (the largest one) of 43 nuclear explosions detonated there between 1948 and 1958.* Had Elugelab, one of Enewetak's islands, had a voice in its selection as the place where Mike would be built, it might have opted out. After Mike's explosion, a crater more than a mile across and 164 feet deep occupied the spot where Elugelab had been.[3]

---

*From 1946 to 1958, there were an additional two dozen tests on Bikini atoll (two hundred miles east of Enewetak and also part of the Marshall Islands). A long history of relocation and resettlement of native people on the islands is a dismal chapter in American history.[2]

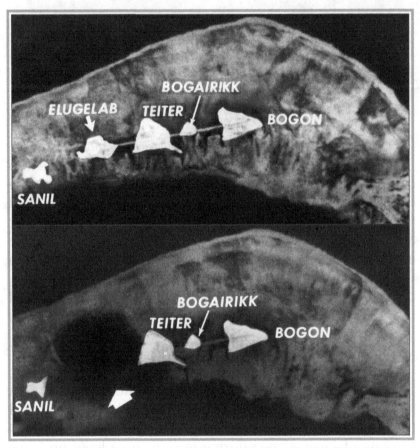

The northern part of the Enewetak atoll, before and after Mike. *Satellite imagery courtesy of Los Alamos National Laboratory Archives.*

To witness Mike, John Wheeler joined Task Force 132, a joint military-civilian operation responsible for preparing, detonating, and measuring the two "shots" that made up Operation Ivy.[4] The first Ivy test was Mike, on which we had worked so hard, fired just after dawn on November 1.* The second was

---

*To be exact, it was at 7:14:59 local time—a hundredth of a second before 7:15 a.m.—on November 1, 1952, or 19:14:59 on October 31 expressed in what was then called Greenwich Mean Time and is now called Universal Time, or Zulu Time.[3]

King, a record-setting fission bomb fired two weeks later. Preparatory work at Enewetak had begun more than six months earlier, in March 1952. The task force, established well before that (on July 8, 1951), moved its headquarters to Parry Island in Enewetak on September 17, 1952, just six weeks before Mike, and closed down that base of operations on November 21, 1952, a little more than a month after King. [4]

A nuclear test, especially in the remote Pacific, involves a lot more than setting up the device and pushing a button. At peak strength, Joint Task Force 132 involved more than 9,000 military personnel and more than 2,000 civilians (not counting those remaining "stateside"); upwards of a hundred water craft, ranging from an aircraft carrier and four destroyers to scores of small boats; and dozens of aircraft, ranging from fighter jets to single-engine four-seaters for island hopping within an atoll (quite like ones I flew much later to tow gliders). Two planes were lost, one of whose pilots was killed. And a lot of money was spent, about $66 million (some $600 million in 2015 dollars). [4]

Wheeler and others who were there just for the spectacle joined Task Force 132.1, the scientific arm of the task force, headed by the physicist Alvin Graves, who, back in Los Alamos, was the leader of Los Alamos's J Division (the test division). Under his "command" (if that's the right term) were more than 2,000 people, a mixture of civilian and military. (The task force had three other large "arms": 132.2, headed by an Army Colonel, with 1,200 people at its peak; 132.3, headed by a Navy Rear Admiral, peaking at more than 5,000 people; and 132.4, headed by an Air Force Brigadier General, reaching some 2,500 people.) Graves's 132.1 was responsible for assembling Mike, handling all of the cryogenics associated with liquid deuterium, and orchestrating the myriad experiments that would monitor its performance. [4]

As the picture on the next page makes clear, Mike, weighing in at 82 tons, [5] looked more like a small factory than

Mike in its construction shed on Elugelab. The vertically mounted "sausage" is on the left, with space left for a fission-bomb trigger above it. The horizontal pipes on the right presumably lead to diagnostic experiments. Mike's total weight has been estimated at 82 tons. *Courtesy of Los Alamos National Laboratory Archives.*

a bomb. The cylinder of deuterium stood vertically, with the fission-bomb trigger at the top (of course not yet there in the picture). Diagnostic experiments (the various ...EX's developed at Los Alamos) were placed near and far, some adjacent to Mike, some as much as two miles distant. [6]

Mike's mushroom cloud reached more than thirty miles into the stratosphere and spread to a width of some 65 miles, even overspreading some of the observers. [5] Those watching, after turning away from the intense light for a few seconds, could see the boiling cloud shooting up and out, and feel its intense heat (arriving as electromagnetic radiation). All in silence. Only after several long minutes did the sound arrive in the form of a shock wave followed by rolling thunder. Harold Agnew had this to say: "Something I will never forget was the heat. Not the blast ... the heat just kept coming on and on. It's really quite a terrifying experience because the heat doesn't go off [as it does on smaller, kiloton shots]." [5]

So far as I know, John Wheeler didn't put his reactions

to the experience down in writing. When he spoke to us about it after his return from the Pacific, the only amazement he expressed was that an entire island could be obliterated. He seemed to take delight in that indicator of Mike's power.* As for me, I remember having the fleeting thought, "It's too bad it worked." But mostly, like my Matterhorn contemporaries, I just felt satisfaction and pleasure that our efforts had contributed to success.

Mike's mushroom cloud. *Courtesy of Los Alamos National Laboratory Archives.*

---

*Gordon Dean, then Chair of the Atomic Energy Commission, also focused on this measure of Mike's muscle when, in briefing President Truman the day after Mike, he reportedly said "The shot island of Elugelab is missing." [7] (Other reports have Dean saying essentially the same thing to President-Elect Eisenhower later in November. [8])

When Mike, in an instant, validated the Teller-Ulam idea and the Garwin design, it was 2:15 p.m. on Friday, October, 31, in Princeton. Those of us waiting for word at Matterhorn got the news of success in the form of an open, unclassified telegram from Wheeler. Whether that came the same afternoon or the next day I can't now remember. What I do remember is that in an excess of caution, Wheeler phrased the news in such Delphic terms that we couldn't tell for sure whether he was announcing success or failure. Since the general tone seemed upbeat, we surmised that the test had been successful. Confirmation came soon after.

Teller, too, used an open telegram to announce the success that he inferred by watching a seismometer in the basement of a geosciences building on the UC Berkeley campus. The explosion took place at 11:15 a.m. that Friday morning in Berkeley, and the seismic waves needed about 15 minutes to travel the more than 4,500 miles from Enewetak to Berkeley.* Teller, by his own account, sat in front of the seismograph in the dark, waited the quarter hour after the scheduled time of the explosion, and then "saw the dot on the seismograph screen do a little dance." He quickly had the seismograph film developed and saw a trace that was "clear, big, and unmistakable," about the magnitude that Teller's geophysicist friend Dave Griggs had predicted. After a short conversation with Ernest Lawrence (inventor of the cyclotron and at that time director of the Berkeley Radiation Lab), Teller composed a three-word telegram, "It's a boy," and sent it off to the physicist Elizabeth Graves (wife of Alvin Graves) at Los Alamos. She and her colleagues at the lab had no trouble interpreting the message. Miraculously, according to Teller, his news of success got to Los Alamos before the classified telegram bearing the official news arrived from Enewetak. [9]

---

*About 4,950 great-circle miles, or 4,600 straight-line miles through the Earth. The speed of the leading wave was about 5 miles per second.

Just after Mike's explosion, its yield was estimated at 10 megatons. This was later refined to 10.4 megatons* (some 700 Hiroshimas), half again as much as we had predicted. [10] One published estimate is that this energy came 23 percent from thermonuclear burning and 77 percent from fission, mostly in the uranium cylinder that housed the deuterium. [11]

Not long after John Wheeler got back to Princeton after the test, he said to me, "Ken, we must have overlooked some energy-generating effect." My response was, "John, given all of the approximations we had to make, and our seriously limited computing power, we were lucky to get within 30 percent of the right answer." (I was counting down from 10, not up from 7.) Now, in retrospect, I have to wonder if my calculations underestimated the yield because we (the Los Alamos and Matterhorn teams combined) underestimated the compression. Perhaps the complex interplay of plasma pressure, ablation pressure, and radiation pressure (see page 157), studied with care in recent years by the independent analyst Cary Sublette, [12] added up to more than we took into account. We will never know. But it remains true that we squeezed remarkably good answers from a computer that today would be considered laughably inadequate.

By the time of the Mike test, the nuclear arms race was well under way. Mike, and H bombs soon to follow, accelerated that race. In August 1953, less than a year after Mike, the Soviet Union exploded a 400-kiloton fission-fusion weapon of "layer-cake" design (see page 5), [13] and followed this up in 1961 with a 50-megaton behemoth† just to show it could be done. [14]) On March 1, 1954 (February 28 in the US), in the Castle Bravo test,

---

*What a tribute to the experimental physicists and engineers that they can figure the energy output of such a maelstrom to 1 percent.

†Which, to this day, holds the record for energy release in a single explosion.

the United States detonated an H bomb fueled with lithium deuteride that yielded 15 megatons. [15] (This device was enriched in Li6 but was still mostly Li7. The Los Alamos designers apparently—and very surprisingly—forgot that Li7, not just Li6, can contribute tritons to add to the thermonuclear burning. The predicted yield, accordingly, was less than half of the actual measured yield. [15] As I discussed on page 160 that extra yield had serious fallout consequences.)

The decades after Mike saw a rapid rise in the world's stockpile of nuclear weapons. Great Britain joined the nuclear club in 1952, France in 1960, China in 1964, India in 1974, Pakistan in 1998, and North Korea in 2006. [16] Israel has so far not publicly acknowledged having nuclear weapons but is believed to have had them since the 1970s. [17] Iran has so far not publicly acknowledged its intent to acquire them. South Africa had a nuclear weapons program that it terminated, [18] and Libya decided against proceeding, even when it had the means within its grasp. [19] Stockpiled nuclear weapons worldwide are estimated to have grown to a maximum of more than 63,000 in the mid-1980s and shrunk gradually after that to a currently estimated 17,000. [20] The number of Hiroshimas still lying in wait remains unimaginably large.

In 2008, in a speech in Berlin, Presidential candidate Barack Obama urged all countries to adopt a policy of nuclear disarmament, with the goal of a world free of nuclear weapons. [21] So far his country (and mine) has not publicly adopted such a policy, nor has any other nuclear power. The only nations to have gone on record in favor of eliminating all nuclear weapons are eight that call themselves the New Agenda Coalition* and that do not themselves possess nuclear weapons. [22]

Zero is indeed the correct target number. Perhaps in your lifetime, young reader. Not in mine.

---

*Brazil, Egypt, Ireland, Mexico, New Zealand, Slovenia, South Africa, and Sweden.

Epilogue

The Vietnam War changed my view of my own country.

Could the United States be trusted to behave intelligently and ethically in the international sphere? I began to doubt. I had joined the H-bomb effort in part because of a conviction that great power in American hands would be used to preserve peace, not to make war—specifically, that the world would be a safer place if the United States got the H bomb before the Soviets did. Now I began to wonder: Is my country, after all, no better than any other? Is it, in a sense, even worse than some because of its willingness to spread death and destruction in the name of preserving freedom?

The anticommunist hysteria of the late 1940s and early 1950s was a sad chapter in American history. It destroyed some careers and was a mighty embarrassment to this country. But it shed no blood. It did not involve the raw exercise of power. I made my own small statements of opposition to that outbreak of insanity by participating in the work of the Scientists Committee on Loyalty Problems and by resisting the signing of loyalty oaths (once by only delaying the signing, once by indeed refusing to sign).

The Vietnam War was different—insanity at a different level. By the mid-1960s some of my academic colleagues and a great many young people across the country had become very audible and visible opponents of the war. My opposition, as I approached my fortieth birthday, was far more muted— at first little more than a show of solidarity by trading in my broadcloth shirt for a blue work shirt and letting my hair grow a little longer. I well remember my first participation in an anti-war demonstration. It was in the fall of 1965, or perhaps the spring of 1966. I joined a group of Quakers lined up along a highway in Costa Mesa, California in the early evening. I stood there holding a burning candle—others had signs as well as candles—as we sought to communicate silently with passing motorists. To a true activist, my participation was a very modest effort indeed. To me, at the time, it seemed like a large step, a daring public display.

Soon thereafter, some of my faculty colleagues at the University of California, Irvine, collected enough contributions to fund one trip to Washington. I was selected as the spokesperson to make the anti-war case to our Representative in Congress. So off I went. James Utt* welcomed me cordially enough in his spacious office and listened politely, after first asking, "Where is the rest of your delegation?" My impact was, I'm sure, close to zero, but perhaps not exactly zero. I had to console myself with that thought.

I made a couple of trips back to the East on other business in the following year, and took advantage of those trips to take part in other public anti-war demonstrations, one in Harrisburg, Pennsylvania, one in Washington, D.C. I made no speeches and carried no signs. I just lent my physical presence to a cause I now believed in, and I felt more comfortable, less "daring," than in that first Costa Mesa vigil.

Oddly, I was once tear-gassed in a demonstration in which I had no part. I was on the Berkeley campus for a meeting—probably in 1967—and was swept up in a horde of mostly students running pell mell from an onslaught of tear gas spread by local or campus police. By that time, causes had been mingled in campus demonstrations. I didn't know whether this one was on behalf of minority rights or women's

---

*Utt, representing the conservative district in Southern California where I lived and where UC Irvine was established, was notable for his consistent opposition to civil rights legislation and to America's membership in the United Nations. (He was, however, politically a bit to the left of his successor, the John Birch Society activist John Schmitz.)

rights, or in opposition to the war or to the practice of grading student performance.

As my very modest activism increased, and my opposition to the Vietnam war grew stronger, I decided that I wanted nothing further to do with weapons work or secret work of any kind. It wasn't that I regretted my participation in the H-bomb program. I took part in it for what seemed to be good reasons at the time, and I never felt afterwards that I had made a "mistake" in joining the effort. As I reported in Chapter 3, I felt that it would be a good thing if the United States acquired an H bomb before the Soviet Union did. I thought of the United States then as a moral nation that could be trusted with weapons of nearly unlimited destructive power and the Soviet Union as a nation that could not be trusted. My faith in the morality of the United States was now seriously fraying, but nothing had happened to make the Soviet Union seem more trustworthy. I remained convinced that it was indeed a good thing that the United States had been the first to achieve a thermonuclear explosion. In the decade-and-a-half since Mike, the world had been a little safer, I felt sure, than it might otherwise have been.

In the summer of 1968 I found myself back in Los Alamos, a wonderful place to live* and a wonderful place to work. This time I was engaged in unclassified research on nuclear structure, no different from my university-based research and free of any secrecy requirements. My decision to do no more weapons work or other secret work had no immediate practical consequences, since I was not being asked to do such work

---

*My last child was born that summer in Los Alamos, in the same hospital where my first child had been born eleven years earlier. As I mentioned in Chapter 7, my wife and I and our brood of children lived that summer in a Bathtub Row house with the Ulams as our next-door neighbors.

that summer and did not expect to be invited to do so in the foreseeable future. Yet I had the gnawing feeling that my decision was potentially reversible if I didn't announce it publicly in some way.

That opportunity came part way through the summer. Los Alamos had a sizable contingent of scientists and others who shared my antiwar sentiments, and we met now and then to talk about what, if anything, we could do to hasten the end of the war. Around mid-summer a regional meeting to address how scientists might have some influence on the decision makers and war-makers in Washington was scheduled for Cloudcroft in southern New Mexico. I attended the small gathering of no more that fifty people and gave a talk. I no longer remember what I had to say or why I was invited to speak. I remember only that I took advantage of the opportunity to state in public that the ongoing war had so strongly affected me that I had decided to do no more weapons work or other classified work. Then I breathed a sigh of relief. With those few words to that small audience, I had made my public statement.

That I have done no weapons or other secret work since then is hardly significant. The last time I did any such work was in the late 1950s, and I have not been invited to do any since. (This reflects more a drying up of the free flow of Defense Department money to the aerospace industry for over-the-horizon projects than it does a judgment about my ability to contribute.) With or without my declaration, I would very likely have remained a non-participant in war work for all the years since Cloudcroft. Yet I felt good about the declaration, and hoped that it would make some other scientists think about how, if at all, their opposition to the Vietnam war should influence their decisions about their own careers.

Acknowledgments

To track down documents and references, I have had the invaluable help of Anthony Eames (my research assistant), Bill Reupke (a Los Alamos alumnus who channeled his keen curiosity on my behalf), and Alex Wellerstein (my rock in the world of physics history). Jeremy Bernstein provided a trove of data and leads, as did Dick Garwin and Carey Sublette. And the trove of all troves, *The Swords of Armegeddon*, was left behind by Chuck Hansen (1947–2003).

Like every author, I have leaned on many people for assistance of one kind or another. My benefactors include Alan Carr and Matthew Hopkins at Los Alamos, Kathy Olesko of George Washington University, and Gino Segrè, my Philadelphia neighbor whose physics writing has served as an inspiration. Martin Levin helped me navigate the tricky waters of information disclosure. Harris Mayer shared with me his recollections of the early days at Los Alamos. Frances Lennie taught me (again) how to prepare an index. Two Mashas have helped with translations: Masha Ford (German) and Masha Spektor (Russian).

It always helps to have friends in high places, and I have been fortunate to have a few: Representative Rush Holt (a physicist as well as a Member of Congress) and his assistant Sarah Steward; Bill Burr at the National Security Archive; Fletcher Whitworth at the Department of Energy; and Terry Fehner, also at DOE. Without their help, some of the once-secret documents that I needed could have been too long delayed to be useful.

Encouragement is important to every author, and I have had a lot of it: from my good physics friend Paul Hewitt (who also initiated the figures in Chapter 6); my wife, Joanne (52 years and counting); and my children Paul Ford, Sarah Ford, Nina Tannenwald, Caroline Richards, Adam Ford, Jason Ford, and Star Ford (balancing their morale-boosting cheers with perceptive critiques). One of my daughters, Nina, has also written about nuclear weapons. My son Adam designed this book.

There are lots of pictures in this book. Rebecca Collinsworth and Savannah Gignac have been of special help in securing many of them. Aimee Slaughter briefed me on Bathtub Row.

It's been a pleasure working with the World Scientific team, notably Jessica Barrows in New Jersey and Alvin Chong in Singapore.

End Notes

Chapter 1 – The Big Idea

1. http://www.nuclearnonproliferation.org/LAMS1225.pdf
2. Rhodes, *Dark Sun*, p. 256.
3. Teller, *Memoirs*, p. 242.
4. Fitzpatrick, "Igniting the Light Elements," p. 126.
5. Holloway, *Stalin and the Bomb*, p. 298.
6. Teller, *Memoirs*, p. 316.
7. Teller interview with Jay Keyworth, Sept. 20, 1979 (Teller "Testament"), pp. 14–15, copy kindly furnished by Richard Garwin; also Teller, *Memoirs*, p. 312.
8. Teller, *Memoirs*.
9. Teller, *Memoirs*, p. 316.
10. Teller, *Memoirs*, p. 314.
11. Teller, *Legacy*, p. 50.
12. Teller, *Memoirs*, p. 316.
13. http://nuclearweaponarchive.org/
14. http://nuclearweaponarchive.org/Usa/Tests/Ivy.html
15. Holloway, *Stalin and the Bomb*, p. 307.
16. Holloway, *Stalin and the Bomb*, p. 196.
17. Goncharov, "American and Soviet," p. 1041.
18. Sakharov, *Memoirs*, p. 182.
19. Goncharov, "American and Soviet," pp. 1038–1041.
20. Holloway, *Stalin and the Bomb*, p. 298.
21. Goncharov, "American and Soviet," p. 1038.
22. http://nuclearweaponarchive.org/Nwfaq/Nfaq8.html#nfaq8.2.2
23. Sakharov, *Memoirs*, p. 102.
24. http://www.atomicarchive.com/History/hbomb/page_08.shtml
25. Holloway, *Stalin and the Bomb*, p. 314.
26. http://fas.org/nuke/guide/usa/nuclear/209chron.pdf

Chapter 2 – The Protagonists

1. Teller, *Memoirs*, p. 299.
2. Shepley and Blair, *The Hydrogen Bomb*.
3. Fitzpatrick, "Igniting the Light Elements," p. 12.
4. Teller, "Many People."
5. Teller, *Memoirs*, footnote on p. 407.
6. Bethe, Hans, Appendix II in the 1989 edition of York, *Advisors*, p. 173.
7. Ibid, pp. 171–172.
8. Teller, *Legacy*, pp. 49–50.
9. Los Alamos National Laboratory Archives, Collection A-1999-019, Box 24, Folder 20.
10. Ulam, *Adventures*, p. 219.
11. Ulam, *Adventures*, p. 311.
12. Teller interview with Jay Keyworth, Sept. 20, 1979 (Teller "Testament"), pp. 14–15, copy kindly furnished by Richard Garwin.
13. Teller, *Memoirs*, pp. 310–311, pp. 314–315.
14. Teller, *Memoirs*, pp. 315–316.
15. Carson Mark, AIP Oral History Interview, http://www.aip.org/history/ohilist/22909.html.

Chapter 3 – The Choice

1. *The Daily Princetonian*, March 6, 1952, p. 1.
2. Princeton University Graduate School archives, Seeley G. Mudd Manuscript Library.
3. Schweber, *Mesons and Fields*.
4. American Physical Society, announcement of prize, http://www.aps.org/units/fhp/newsletters/fall2010/pais.cfm
5. Schweber, *Nuclear Forces*.
6. Harvard University Press, reviews of Schweber book, http://www.hup.harvard.edu/catalog.php?isbn=9780674065871&content=reviews
7. Wheeler, *Geons*, p. 183.
8. Wheeler, *Geons*, p. 183.
9. Wheeler, *Geons*, p. 183.
10. Wheeler, *Geons*, p. 186.
11. Wheeler, *Geons*, p. 184.
12. Rainwater, James. "Nuclear Energy Level Argument for a Spheroidal Nuclear Model," *Physical Review* 79, 432 (1950).

13. Rainwater's Nobel autobiography, http://www.nobelprize.org/nobel_prizes/physics/laureates/1975/rainwater-autobio.html
14. Wheeler, *Geons*, p. 187.
15. Wheeler, *Geons*, p. 187.
16. Wheeler, *Geons*, p. 188.
17. Wheeler, *Geons*, p. 189.
18. Telegram kindly provided by John Wheeler's daughter Letitia Wheeler Ufford.
19. Wheeler, *Geons*, p. 189.
20. Teller, *Memoirs*, p. 295.
21. *New York Times*, Feb. 1, 1950, p.1.
22. See, for example, the commentary of Alex Wallerstein, http://nuclearsecrecy.com/blog/2012/06/18/what-if-truman-didnt-order-h-bomb-crash-program/
23. Wheeler, *Geons*, p. 191.
24. Wheeler, *Geons*, p. 189.
25. Wheeler, *Geons*, p. 191.
26. Carson Mark, AIP Oral History interview, http://www.aip.org/history/ohilist/22909.html.
27. Wheeler, *Geons*, p. 192.
28. Personal communication, Letitia Wheeler Ufford.
29. Wheeler, *Geons*, p. 190.

Chapter 4 – The Scientists, the Officials, and the President

1. Part II of the report cited in end note 3.
2. Hewlett and Duncan, *Atomic Shield*, p. 665.
3. http://www.pbs.org/wgbh/amex/bomb/filmmore/reference/primary/extractsofgeneral.html
4. Hewlett and Duncan, *Atomic Shield*, pp. 545–46.
5. McCullough, *Truman*, p. 756.
6. Hewlett and Duncan, *Atomic Shield*, p. 664.
7. http://www.nuclearfiles.org/menu/library/correspondence/strauss-lewis/corr_strauss_1949-11-25.htm
8. McCullough, *Truman*, p. 761.
9. http://www.nuclearfiles.org/menu/library/correspondence/truman-harry/corr_truman_1950-01-31.htm
10. McCullough, *Truman*, p. 761.

11. Potter, Philip, "Decision," *Baltimore Sun*, February 1, 1950, p. 1.
12. McCullough, *Truman*, p. 757.
13. McCullough, *Truman*, pp. 751–754.
14. McCullough, *Truman*, p. 762.
15. McCullough, *Truman*, pp. 762–763.
16. Hewlett and Duncan, *Atomic Shield*, p. 408.
17. Teller, *Memoirs*, p. 286.
18. Teller, *Memoirs*, p. 283.
19. Teller, *Memoirs*, p. 283.
20. Teller, *Memoirs*, p. 286.
21. Teller, *Memoirs*, pp. 287–288.
22. Teller, *Memoirs*, p. 290

Chapter 5 – Nuclear Energy

1. Romer, *Restless Atom*, p. 26. Romer's book is not a primary source, but it is carefully researched and, I believe, generally accurate.
2. Romer, *Restless Atom*, p. 28.
3. Romer, *Restless Atom*, p. 24.
4. Romer, *Restless Atom*, p. 27.
5. Romer, *Restless Atom*, p. 30.
6. http://www.etymonline.com/index.php?term=radioactive
7. http://en.wikipedia.org/wiki/Paul_Ulrich_Villard.
8. Rutherford, *Radio-Activity*.
9. Einstein, Albert, "Does the Inertia of a Body Depend on its Energy Content?," *Annalen der Physik* **18**, 639–641 (1905).
10. Thomson, J. J., "On the Structure of the Atom," Philosophical Magazine Series 6, 7, 237 (1904).
11. Heilbron, *Rutherford*, p. 57.
12. Heilbron, *Rutherford*, p. 65.
13. Rutherford, Ernest, "The Scattering of α and β Particles by Matter and the Structure of the Atom," *Philosophical Magazine* Series 6, 21, 669–688 (1911).
14. www.physics.ox.ac.uk/documents/PUS/dis/notebook.htm
15. Andrade, *Rutherford*, p. 111.
16. Wells, *World Set Free*.
17. Chadwick, James, "The Possible Existence of a Neutron," *Proceedings of the Royal Society of London*, **A136**, 692 (1932).

18. Cockcroft, J. D., and E. T. S. Walton, "Experiments with High Velocity Positive Ions. II — The Disintegration of Elements by High Velocity Protons," *Proceedings of the Royal Society* A, 137, 229–242 (1932).
19. Szilard, *Collected Works*, p. 530.
20. *The Times of London*, September 12, 1933.
21. Szilard, *Collected Works*, p. 529.
22. Weart and G. W. Szilard, *Leo Szilard*, p. 53.
23. Weart and G. W. Szilard, *Leo Szilard*, p. 60.
24. Rhodes, *Making of Atomic Bomb*, p. 67.
25. Rhodes, *Making of Atomic Bomb*, pp. 251–262.
26. von Weizsäcker, C. F., *Zeitschrift fur Physik* **96**, 431–458 (1935).
27. Frisch, *What Little I Remember*, p. 116.
28. Wheeler, *Geons*, p. 21.
29. Frisch, *What Little I Remember*, p. 116.
30. Meitner, L. and O. Frisch, *Nature* **143**, 239 (February 11, 1939).
31. Wheeler, *Geons*, p. 17.
32. http://en.wikipedia.org/wiki/Francis_William_Aston
33. http://en.wikipedia.org/wiki/History_of_mass_spectrometry
34. Eddington, *New Pathways*, p. 167.
35. Schweber, *Nuclear Forces*, pp. 346–352.

Chapter 7 – Going West

1. Miller, *U.S. Navy*, p. 168.
2. Conant, 109 *East Palace*, p. 57; Wheeler, *Geons*, p. 195.

Chapter 8 – A New World

1. http://sunsite.berkeley.edu/uchistory/archives_exhibits/loyaltyoath/regent_resolution.html
2. http://en.wikipedia.org/wiki/David_S._Saxon
3. http://sunsite.berkeley.edu/uchistory/archives_exhibits/loyaltyoath/oaths.html
4. http://sunsite.berkeley.edu/uchistory/archives_exhibits/loyaltyoath/resolution.html
5. Tyler, Carroll L., "Report of the Manager, Santa Fe Operations, U. S. Atomic Energy Commission, July 1950 to January 1954," p. 63. https://www.osti.gov/opennet/index.jsp Accession number NV0079010.

6. Teller, *Memoirs*, p. 297.

7. Letter, Norris Bradbury to Carroll Tyler, February 28, 1950, Chuck Hansen Files, Box 13, National Security Archive, Washington D.C.

8. Memo, US Atomic Energy Commission to its Santa Fe Operations Office, March 3, 1950, Chuck Hansen Files, Box 13, National Security Archive, Washington D.C.

9. http://www.losalamoshistory.org/school.htm

10. Wheeler, *Geons*, p. 160.

11. Ulam, *Adventures*, p. 214.

12. Schweber, *Shadow*, p. 165.

13. Bethe, Hans, Appendix II in the 1989 edition of York, *Advisors*.

14. Ulam, *Adventures*, p. 66.

15. Ulam, *Adventures*, p. 66.

16. Ulam, *Adventures*, p. 136.

17. Ulam, *Adventures*, p. 212.

18. von Neumann, John, and Klaus E.J. Fuchs, "Improvements in methods and means for utilizing nuclear energy," Record of Invention, Case No. S-5292X, copy in Records of the Joint Committee on Atomic Energy, RG 128, Declassified Materials from Classified Boxes, Series 2: General Subject Files, Box 59, File #2910, "Thermonuclear Program." See http://blog.nuclearsecrecy.com/wp-content/uploads/2013/08/1946-Fuchs-von-Neumann-patent-record-of-invention.pdf.

19. Bernstein, "von Neumann and Fuchs."

20. Herken, *Brotherhood of the Bomb*, p. 171.

21. USSR Atomic Project, Vol. III, "Hydrogen Bomb 1945–1956, Book 1." General editor L.D. Riabev, compilation by G. A. Goncharov and P. P. Maksimenko. State Corporation of Atomic Energy, Moscow-Sarov 2008. This document is on file at the Niels Bohr Library and Archives of the American Institute of Physics, College Park, Maryland. According to Norman Dombey (private communication) the notes on Fermi's 1945 lectures that appear in this Russian document were taken not by Fuchs but by the British physicist Philip Burton Moon.

22. Bretscher, et al., "Report of Conference on the Super," (16 February 1950), LA-575-DEL, Los Alamos National Laboratory, online at: http://library.lanl.gov/cgi-bin/getfile?00407902.pdf.

23. Memo Robert Lamphere to Herbert Hoover, June 6, 1950, serial 1412, Klaus Fuchs file, no. 65-58805, FBI. See Herken, *Brotherhood of the Bomb*, p. 374, note 92.

Chapter 9 – The Classical Super

1. http://en.wikipedia.org/wiki/Space_Songs; also http://www.last.fm/music/Tom+Glazer+&+Dottie+Evans/Space+Songs
2. Rhodes, *Making of Atomic Bomb*, p. 415.
3. Serber, *Primer*.
4. Serber, *Serber Says*.
5. Teller, *Memoirs*, p. 157.
6. Teller, *Memoirs*, p. 157.
7. Teller, *Memoirs*, p. 159.
8. Fitzpatrick, "Igniting the Light Elements," p. 59.
9. http://www.atomicarchive.com/History/firstpile/firstpile_05.shtml
10. http://www.mphpa.org/classic/HISTORY/H-05.htm
11. http://www.mphpa.org/classic/HISTORY/H-06c1-2.htm
12. Teller, *Memoirs*, p. 177.
13. Teller, "Many People," p. 269.
14. Hargittai, *Judging Edward Teller*, p. 143.
15. http://www.aip.org/history/acap/institutions/manhattan.jsp
16. See Note 17 below and also Chapter 8 Note 21.
17. "Material No. 713a, 'Atomic superbomb,'" in L.D. Riabev, ed., USSR Atomic Project: Documents and Materials, Vol. 3, Book 1, document no. 31, pp. 93–108. See http://blog.nuclearsecrecy.com/wp-content/uploads/2013/08/APSSSR-T3K1-Material-713.pdf.
18. Bernstein, "von Neumann and Fuchs," p. 45.
19. Lothar Nordheim, AIP Oral History interview, http://www.aip.org/history/ohilist/5074.html.
20. Galison, *Image and Logic*, p. 609; Goldstine, *The Computer*, p. 240.
21. Ulam, *Adventures*, p. 214.
22. http://www.srs.gov/general/about/history1.htm
23. http://energy.gov/em/savannah-river-site
24. Teller, *Legacy*, p. 48.

Chapter 10 – Calculating and Testing

1. Letter from Robert Oppenheimer to Norris Bradbury, July 19, 1950. Nuclear Testing Archive, Las Vegas, Nevada, document NV 0071727.
2. Letter from Robert Oppenheimer to Norris Bradbury, July 19, 1950. Nuclear Testing Archive, Las Vegas, Nevada, document NV 0071726.
3. Teller, Memoirs, p. 279.

4. Hansen, *Swords of Armageddon*, Vol. 2, p. 153.
5. Hansen, *Swords of Armageddon*, Vol. 2, p. 154.
6. Hansen, *Swords of Armageddon*, Vol. 2, p. 154.
7. http://en.wikipedia.org/wiki/Ivy_Mike
8. http://en.wikipedia.org/wiki/Historical_nuclear_weapons_stockpiles_and_nuclear_tests_by_country
9. http://en.wikipedia.org/wiki/List_of_nuclear_weapons_tests_of_the_United_States
10. http://en.wikipedia.org/wiki/Operation_Sandstone#Organization
11. http://en.wikipedia.org/wiki/Nevada_National_Security_Site
12. Fradkin, *Fallout*.
13. Weart, *Nuclear Fear*, p. 98.
14. Bradbury, Norris, memo to Edward Teller, March 15, 1950. Los Alamos National Laboratory Archives, Collection A-1999-019, Box 24, Folder 9.
15. http://www.nnsa.energy.gov/aboutus/ouroperations/generalcounsel/foia/reading-room-after2000?page=9
16. http://www.nnsa.energy.gov/sites/default/files/nnsa/foiareadingroom/RR00232.pdf
17. Hansen, *Armageddon*, Vol. II, pp. 249–257.
18. Hansen, *Armageddon*, Vol. II, p. 253.
19. Herken, *Brotherhood*, p. 154.
20. Hansen, *Armageddon*, Vol. II, pp. 257–260.

Chapter 11 – Constructing Matterhorn

1. Letter of January 8, 1951 John Wheeler to Allen Shenstone mentioning possible new project. Princeton University Seeley G. Mudd Manuscript Library.
2. Memo of January 31, 1951 Allen Shenstone to President Dodds opposing new project and citing Oppenheimer's agreement with his position. Princeton University Seeley G. Mudd Manuscript Library.
3. Letter of January 30, 1951 Allen Shenstone to Henry de Wolf Smyth opposing new project. Princeton University Seeley G. Mudd Manuscript Library.
4. Letter of February 20, 1951 Don Hamilton to President Dodds supporting new project. Princeton University Seeley G. Mudd Manuscript Library.
5. Letter of April 11, 1951 Don Hamilton to John Wheeler re salary structure for new project. Princeton University Seeley G. Mudd Manuscript Library.

6. McCullough, *Truman*, p. 833.
7. Wheeler, *Geons*, p. 39.
8. Misner, Thorne, and Wheeler, *Gravitation*.
9. *Bulletin of the American Physical Society*, vol. 25, Dec. 28, 1950, pp. 12–13.
10. This is the author's recollection.
11. Lyman Spitzer, AIP Oral History interview, http://www.aip.org/history/ohilist/4901_1.html
12. Wheeler, *Geons*, p. 216.
13. Letter of December 3, 1951 Robert Kirkman to Roy Woodrow forbidding discussions of secret work between staff members of Matterhorn B and S. Princeton University Seeley G. Mudd Manuscript Library.
14. Letter of April 13, 1951 Don Hamilton to John Wheeler on salaries and other matters for new project. Princeton University Seeley G. Mudd Manuscript Library.
15. Memo of March 7, 1951, Roy Woodrow to Lyman Spitzer and others on project oversight committee. Princeton University Seeley G. Mudd Manuscript Library.
16. Letter of April 6, 1951 John Wheeler to Wendell Furry recruiting for new project, Princeton University Seeley G. Mudd Manuscript Library.
17. Wheeler, *Geons*, p. 218.
18. Wheeler, *Geons*, p. 219.
19. Wheeler, *Geons*, p. 218.
20. Spitzer letter to Rolf Sinclair, May 9, 1994. Copy available from author.
21. http://web.ornl.gov/sci/fed/stelnews/
22. Merritt, *Forrestal*.

Chapter 12 – Academia Cowers

1. Cole, *Something Incredibly Wonderful*.
2. Rossi Lomanitz, AIP Oral History interview, http://www.aip.org/history/ohilist/24703_3.html
3. Wheeler, *Geons*, p. 216.
4. http://en.wikipedia.org/wiki/David_Bohm. See also Peat, *Infinite Potential*, pp. 62–63.
5. David Bohm, AIP Oral History interview, http://www.aip.org/history/ohilist/32977_3.html
6. Wheeler, *Geons*, p. 216.
7. Peat, *Infinite Potential*, pp. 91–93.

8. Peat, *Infinite Potential*, p. 98.
9. Peat, *Infinite Potential*, p. 99.
10. David Bohm, AIP Oral History interview, http://www.aip.org/history/ohilist/32977_4.html.
11. Wang, *American Science*, p. 213. See also "The First Year of the SCLP," *Science* **111**, 220–225 and 238 (1950).
12. List of Sponsors, American Institute of Physics Niels Bohr Library and Archives, College Park, Maryland, Samuel A. Goudsmit collection, Box 40, Folder 3a.
13. Wang, *American Science*, p. 220.
14. Wang, *American Science*, p. 220.
15. Letter from Lyman Spitzer to SCLP Sponsors, American Institute of Physics Niels Bohr Library and Archives, College Park, Maryland, Samuel A. Goudsmit collection, Box 40, Folder 3a.
16. Minutes of June 3, 1950 SCLP meeting, American Institute of Physics Niels Bohr Library and Archives, College Park, Maryland, Samuel A. Goudsmit collection, Box 40, Folder 3a.
17. Ford, K. W. and D. Bohm, "Nuclear Size Resonances," *Physical Review* **79**, 745–746 (1950).

Chapter 13 – New Mexico, New York, and New Jersey

1. Wheeler, *Geons*, p. 218.
2. Teller, E., and J. Wheeler, LAMD-443, August 1950. See also Hansen, *Swords of Armageddon*, Vol. 2, p. 153.
3. Hansen, *Swords of Armageddon*, Vol. 2, p. 154.
4. Wheeler, *Geons*, p. 207; also Hewlett, *Atomic Shield*, p. 527.
5. Hansen, *Swords of Armageddon*, Vol. II, p. 253.
6. Hansen, *Swords of Armageddon*, Vol. II, pp. 257–258.
7. http://en.wikipedia.org/wiki/IBM_CPC
8. http://ed-thelen.org/comp-hist/xxx-ibm-cpc.html

Chapter 14 – The Garwin Design

1. Hewlett and Duncan, *Atomic Shield*, p. 664.

2. Galison and Bernstein, *In Any Light*, p. 323; Hansen, *Swords of Armageddon*, Vol. 2, p. 264; Teller, *Memoirs*, p. 324; Hewlett and Duncan, *Atomic Shield*, Vol. II, p. 544.

3. U. S. AEC, *Transcript*, p. 81.

4. U. S. AEC, *Transcript*, p. 251.

5. Teller, Edward, memo of Oct. 10, 1949 to Norris Bradbury, Los Alamos National Laboratory Archives, Collection A-1999-019, Box 207, Folder 28.

6. Garwin AIP interview, about 20 percent of the way through the interview.

7. Garwin, Richard, "Some Preliminary Indications of the Shape and Construction of a Sausage, Based on Ideas Prevailing in July 1951," Los Alamos Scientific Laboratory. report LAMD-746.

8. http://www.nnsa.energy.gov/aboutus/ouroperations/generalcounsel/foia/reading-room-after2000?

9. Garwin AIP interview, about 35 percent of the way through the interview.

10. Teller, Edward, "The Sausage: A New Thermonuclear System," LA-1230, Los Alamos Scientific Laboratory, April 4, 1951.

11. Garwin AIP interview, about 40 percent of the way through the interview.

12. Rhodes, *Dark Sun*, p. 493; also Hansen, *Armageddon*, vol. 2, pp. 47 and 278.

13. http://en.wikipedia.org/wiki/Ivy_Mike#Detonation

14. Garwin AIP interview, about 45 percent of the way through the interview.

15. http://en.wikipedia.org/wiki/Mark_16_nuclear_bomb

16. http://en.wikipedia.org/wiki/Castle_Bravo

17. Hewlett, Holl, and Anders, *Atoms for Peace and War*, pp. 175–177.

Chapter 15 – Climbing Matterhorn

1. Dyson, *Turing's Cathedral*, p. 274.

2. http://en.wikipedia.org/wiki/Williams_tube

3. http://en.wikipedia.org/wiki/SEAC_(computer)

4. http://en.wikipedia.org/wiki/Marshall_Rosenbluth

5. http://www.theguardian.com/news/2003/oct/04/guardianobituaries.highereducation1

6. Wheeler, "Matterhorn."

7. Wheeler, *Geons*, p. 218.

8. Wheeler, "Matterhorn," p. 58.
9. Wheeler, "Matterhorn," p. 58.
10. Feynman, *Letters of Feynman*, pp. 82–85.
11. Wheeler, "Matterhorn," p. 61.
12. http://nuclearweaponarchive.org/Nwfaq/Nfaq8.html
13. Rhodes, *Dark Sun*, p. 503.
14. http://en.wikipedia.org/wiki/Mark_5_nuclear_bomb
15. Marshall Rosenbluth oral history interview, http://www.aip.org/history/ohilist/28636_1.html, about 45 percent of the way through the interview.

Chapter 16 – More Than a Boy

1. http://en.wikipedia.org/wiki/Enewetak_Atoll
2. http://en.wikipedia.org/wiki/Bikini_Atoll#Residents_relocated. See also http://www-ns.iaea.org/appraisals/bikini-atoll.asp.
3. http://nuclearweaponarchive.org/Usa/Tests/Ivy.html
4. Joint Task Force 132 Final Report, http://www.dtra.mil/documents/ntpr/historical/DNA6036F.pdf
5. http://www.ctbto.org/specials/testing-times/1-november-1952-ivy-mike/?textonly=1
6. http://en.wikipedia.org/wiki/Ivy_Mike
7. MacKenzie, *Inventing Acccuracy*, p. 106.
8. http://www.army.mil/article/47341/The_Island_is_Missing__/. See also Herken, *Brotherhood*, p. 262.
9. Teller, *Memoirs*, p. 352.
10. http://en.wikipedia.org/wiki/Operation_Ivy
11. http://nuclearweaponarchive.org/Usa/Tests/Ivy.html
12. http://nuclearweaponarchive.org/Library/Teller.html
13. Holloway, *Stalin and the Bomb*, p. 307.
14. http://en.wikipedia.org/wiki/Tsar_Bomba
15. http://nuclearweaponarchive.org/Usa/Tests/Castle.html
16. http://en.wikipedia.org/wiki/List_of_states_with_nuclear_weapons
17. http://www.nti.org/country-profiles/israel/nuclear/
18. http://en.wikipedia.org/wiki/South_Africa_and_weapons_of_mass_destruction#Testing_the_first_device
19. http://en.wikipedia.org/wiki/Abdul_Qadeer_Khan#North_Korea.2C_Iran_and_Libya

20. http://en.wikipedia.org/wiki/Historical_nuclear_weapons_stockpiles_and_nuclear_tests_by_country
21. http://edition.cnn.com/2008/POLITICS/07/24/obama.words/
22. http://en.wikipedia.org/wiki/Nuclear_Free_World_Policy

Bibliography

Andrade, E. N. da C., *Rutherford and the Nature of the Atom*. New York: Doubleday (1964).

Bernstein, Jeremy. "John von Neumann and Klaus Fuchs: An Unlikely Collaboration," *Physics in Perspective*, vol. 12, No. 1, pp. 36–50 (2010).

Birks, J. B. (Ed.). *Rutherford at Manchester*. New York: W. A. Benjamin (1963).

Conant, Jennet, *109 East Palace: Robert Oppenheimer and the Secret City of Los Alamos*. New York: Simon and Schuster (2006).

Dyson, George, *Turing's Cathedral: The Origins of the Digital Universe*. New York: Pantheon (2012).

Eddington, Arthur. *New Pathways in Science: Messenger Lectures (1934)*. Cambridge, England: Cambridge University Press (1935).

Feynman, Michelle. ed. *Perfectly Reasonable Deviations from the Beaten Track: The Letters of Richard P. Feynman*. New York: Basic Books (2005).

Fitzpatrick, Anne. "Igniting the Light Elements: The Los Alamos Thermonuclear Weapon Project, 1942–1952." Los Alamos National Laboratory, Report LA-13577-T (July, 1999).

Fredkin, Philip L. *Fallout: An American Nuclear Tragedy*. Boulder, Colorado: Johnson Books (2004).

Frisch, Otto, *What Little I Remember*. Cambridge, England: Cambridge University Press (1979).

Galison, Peter, *Image and Logic: A Material Culture of Microphysics*. Chicago: University of Chicago Press (1997).

Galison, Peter, and Barton Bernstein. "In Any Light: Scientists and the Decision to Build the Superbomb, 1952–1954," *Historical Studies in the Physical and Biological Sciences*, Vol. 19, No. 2 (1989), pp. 267– 347.

Garwin, Richard, AIP oral history interview, http://www.aip.org/history/ohilist/35680.html.

Goldstine, Herman H., *The Computer from Pascal to von Neumann*. Princeton: Princeton University Press (1980).

Goncharov, G. A. "American and Soviet H-bomb development programmes: historical background." *Physics-Uspekhi* **39** (10), 1033–1044 (1996).

Hansen, Chuck. *The Swords of Armegeddon, Version 2.* (4 volumes) Sunnyvale, Calif.: Chukelea Publications (2007).

Hargittai, Istvan. *Judging Edward Teller: A Closer Look at One of the Most Influential Scientists of the Twentieth Century.* Amherst, N.Y.: Prometheus (2010).

Heilbron, John L. *Ernest Rutherford: And the Explosion of Atoms.* Oxford: Oxford University Press (2003).

Herken, Gregg. *Brotherhood of the Bomb: The Tangled Lives and Loyalties of Robert Oppenheimer, Ernest Lawrence, and Edward Teller.* New York: Henry Holt (2002).

Hewlett, Richard G., and Francis Duncan. *Atomic Shield, 1947/1952* (Vol. II of *A History of the United States Atomic Energy Commission*). University Park, Pennsylvania: The Pennsylvania State University Press (1969).

Hewlett, Richard G., Jack M. Holl, and Roger M. Anders. *Atoms for Peace and War 1953–1961* (Vol. III of *A History of the United States Atomic Energy Commission*). Berkeley, California: University of California Press (1989).

Holloway, David. *Stalin and the Bomb: The Soviet Union and Atomic Energy, 1939–1956.* New Haven: Yale University Press (1994).

MacKenzie, Donald A. *Inventing Accuracy: A Historical Sociology of Nuclear Missile Guidance.* Cambridge, Massachusetts: MIT Press (1993).

Mark, J. Carson. Oral History Transcript, Feb. 24, 1995, AIP Niels Bohr Library and Archives, http://www.aip.org/history/ohilist/22909.html.

McCullough, David. *Truman.* New York: Simon and Schuster (1992).

Merritt, J. I. *Princeton's James Forrestal Campus: 50 Years of Sponsored Research.* Princeton: Trustees of Princeton University (2002).

Miller, Nathan. *U.S. Navy: A History.* Annapolis, Maryland: Naval Institute Press, (1997).

Misner, Charles, Kip Thorne, and John Wheeler. *Gravitation.* San Francisco: W. H. Freeman (1973).

Peat, F. David. *Infinite Potential: The Life and Times of David Bohm.* Reading, Massachusetts: Addison-Wesley (1997).

Rhodes, Richard. *Dark Sun: The Making of the Hydrogen Bomb.* New York: Simon and Schuster (1995).

Rhodes, Richard. *The Making of the Atomic Bomb.* New York: Simon and Schuster (1986).

Romer, Alfred. *The Restless Atom.* New York: Dover (1982).

Rutherford, Ernest. *Radio-Activity*. Cambridge, England: Cambridge University Press (1904).

Sakharov, Andrei. *Memoirs*. New York: Vintage Books (1992).

Schweber, Silvan S., Hans Bethe, and Frederic de Hoffmann. *Mesons and Fields*. Evanston, Illinois: Row Peterson (1955).

Schweber, Silvan S. *In the Shadow of the Bomb: Oppenheimer, Bethe, and the Moral Responsibility of the Scientist*. Princeton: Princeton University Press (2000).

Schweber, Silvan S. *Nuclear Forces: The Making of the Physicist Hans Bethe*. Cambridge, Massachusetts: Harvard University Press (2012).

Serber, Robert. *Los Alamos Primer*. Berkeley, California: University of California Press (1992). (Based on lectures first delivered at Los Alamos in April 1943.)

Serber, Robert. *Serber Says: About Nuclear Physics*. Singapore: World Scientific (1987). (Updated version of lecture notes from 1947.)

Shepley, James R., and Clay Blair, Jr. *The Hydrogen Bomb: The Men, the Menace, the Mechanism*. New York: David McKay (1954).

Szilard, Leo. *The Collected Works: Scientific Papers*. Cambridge, Massachusetts: MIT Press (1972).

Teller, Edward. "The Work of Many People," *Science* **121**, 267–275 (February 25, 1955).

Teller, Edward, with Allen Brown. *The Legacy of Hiroshima*. Garden City, New York: Doubleday (1962).

Teller, Edward, with Judith Shoolery. *Memoirs. A Twentieth-Century Journey in Science and Politics*. Berkeley: Perseus Publishing (2001).

Ulam, Stanislaw. *Adventures of a Mathematician*. New York: Charles Scribner's Sons (1976).

United States Atomic Energy Commission. *In the Matter of J. Robert Oppenheimer: Transcript of hearing before personnel security board and texts of principal documents and letters*. Washington: U.S. Government Printing Office (1954). (Also MIT Press, 1971.)

Wang, Jessica. *American Science in an Age of Anxiety: Scientists, Anticommunism, and the Cold War*. Chapel Hill, North Carolina: University of North Carolina Press (1998).

Weart, Spencer. *The Rise of Nuclear Fear*. Cambridge, Massachusetts: Harvard University Press (2012).

Weart, Spencer, and Gertrud Weiss Szilard (Eds.). *Leo Szilard: His Version of the Facts*. Cambridge, Massachusetts: MIT Press (1978).

Wellerstein, Alex. "Restricted Data: The Nuclear Secrecy Blog." blog.nuclearsecrecy.com.

Wheeler, John Archibald, with Kenneth Ford. *Geons, Black Holes, and Quantum Foam: A Life in Physics*. New York: W. W. Norton (1998).

Wheeler, John Archibald, with Ford, Frieman, McIntosh, and Noyes. "Project Matterhorn Final Report" (31 August 1953), PM-B-37. Secured through a FOIA request to the U.S. Department of Energy. Copy available from author.

Wells, H. G. *The World Set Free*. New York: E. P. Dutton (1914).

York, Herbert. *The Advisors: Oppenheimer, Teller, and the Superbomb*. San Francisco: W. H. Freeman (1976); and Stanford, California: Stanford University Press (1989).

Index

An italic page number refers to an illustration on that page. A page number followed by the letter n refers to a footnote on that page. The End Notes and Bibliography are not indexed. The author is not indexed except for two pictures in which he appears.

109 East Plaza Avenue, 75, 76
1300 20th Street, 77, 78

Abelson, Philip (Phil), 55
Aberdeen, Maryland, 103
ablation, 9, 157
accelerator, 65
Acheson, Dean, 42
Admiralty, British, 52, 53n
AEC, 6
    1949 composition of, 41n
    and formation of Matterhorn, 121
    forwarding GAC report to, 40
    June 1951 advisory committee to, 152–153
    purchase of SEAC time by, 163
    and question of thermonuclear priority, 30
    and Seaborg, 15
aerospace industry, 188
Agnew, Harold, 175
Agronsky, Martin, 79
Aiken, South Carolina, 104, 113
Alamogordo, New Mexico, 112
"alarm clock," 5, 10, 106, 146
Alferov, Zhores, 11n
alpha particles, 48, 49, 50, 63
scattering of, 49
alpha rays, 46
Alsos mission, 137n
Alverez, Luis, 55

American Physical Society. See APS
anticommunism, 185
antineutrinos, 104
anti-war demonstrations, 185, 186
APS, 123
    Los Angeles meeting of, 20, 123
    New York Meeting of, 123
    west coast meetings of, 123–124
Arnold, William, 54n
Aron, Walter, 141
arsenals. See stockpiles
Ashley Pond, 85
Aspen, Colorado, 126
Aston, Francis William, 55
Atomic Energy Commission. See AEC

B-36 bomber, 159
bachelor officer quarters, 151
barium, 53
Bathtub Row, 120, 123, 187n
Becquerel, Henri, 44–45
Berger, Jay, 141
Bergmann, Peter, 123
Berkeley, 182, 186
    1942 conference in, 82, 93
Berlin Blockade, 122
Bernstein, Jeremy, 90, 189
beryllium oxide, 100, 101
beta particles, 50
beta rays, 46

Bethe, Hans, 25, 58
  and 1942 conference, 93n, 96
  1949 meeting with Teller, 43
  1951 interaction with Garwin,
    155
  at June 1951 meeting, 152
  long life of, 90
  as Los Alamos consultant, 87,
    88, 119
  and stellar energy, 57, 93
  and Teller-Ulam idea, 14
  as wartime T-Division head, 97
Bethe, Rose, 96
Beverly Hills, California, 125
Bikini atoll, 11, 112, 177n
binding energy, 63
  per nucleon, 64, 66
Birkbeck College, 135
bismuth, 62, 63
Black Mesa, 78
Blair, Clay, 14n
Bloch, Felix, 93n
Bochner, Solomon, 141
Bohm, David, 133–135
  acquittal of, 135
  arrest of, 134
  friendship with author, 136, 137
  joint paper with author, 138
  loss of Princeton job, 139
  and SCLP, 136
Bohr, Aage, 28n, 29n
Bohr, Niels, 28, 29, 54, 55
  and news of fission, 93
  and Soviet threat, 32
"bomb in a box," 22, 23
boosting, 5, 118, 148
Bradbury, Norris, 17n, 18, 19, 152
  Bethe's link to, 88
  his endorsement of
    Matterhorn, 126
  and Family Committee, 114–115
  his interaction with Teller, 19
  and loyalty oath, 82–83
  and recruitment, 119

  and September 1950 GAC
    meeting, 106
  and Teller, 21
  and Ulam, 16, 17
Bravo test, 160, 168, 183–184
Brazil, 135, 184n
Broadway Limited, 143, 144n
Buckley, Oliver E., 37n, 38
Budapest, 156
Burr, William (Bill), 189
Buster Jangle, 113

Caldwell, David, 85
California, State of, 81–82
California, University of. See
  University of California
card-programmed calculators. See
  CPCs
carolinium, 50, 60
Carr, Alan, 189
Carryall. See Chevrolet Carryall
Carter, David, 25, 141
Casey, Roberta, 141
Castle Bravo test. See Bravo test
Cather, Willa, 76
Cerrillos Road, 75
Chaco Canyon, 84
Chevrolet Carryall, 131, 165
  acquisition of, 73
  final fate of, 162
  misadventures with, 161, 162
  trip west in, 73–74
  western trips in, 84, 161
Chief (calculations), 173
China in nuclear club, 184
Churchill, Winston, 122
classical Super. See Super,
  classical
Clendenin, William, 141
Cleveland, Ohio, 131
Clines Corners, 74
Cloudcroft, New Mexico, 188
Cockcroft, John, 50
Cockcroft-Walton experiment, 51,
  52, 56

Cold War, 122
collective model, 28
Collinsworth, Rebecca, 190
Columbia University, 55, 93
compression, 6, 9, 21
    reported in PM-B-37, 172
computers
    as people, 5n, 84, 106n, 144
    stored-program, 149
    See also specific computers
"computresses," 5n
Conant, James B., 37n, 38
    his letter opposing H-bomb
        program, 43
Copenhagen, 27, 32, 54
Cornell University, 57, 87
cosmic background radiation, 67
Costa Mesa, California, 186
Cowan, Clyde, 104
CPCs, 5n, 103
    in New Mexico, 105, 144
    in New York City, 144, 151–152
"crash program," 31, 122
Crossroads, 112
Curie, Marie, 45, 45–46
    and radioactive coinage, 45
Curie, Pierre, 46

DD reactions, 109–110, 157
Dean, Gordon, 41, 106, 152n, 181n
Death Comes for the Archbishop, 76
Defense Department, 188
de Hoffmann, Frederic (Freddie),
    15, 16, 18, 20, 87
    April 1951 report by, 156
    meeting Wheeler in Nice,
        32–33, 34
deuterium
    in Gedankenexperiment, 68–70
    as thermonuclear fuel, 95, 96,
        104, 111
deuteron, 63
Devaney, Joseph (Joe), 85
dewar, 156
differential equations, 145

diffusion of neutrons, 94
disarmament, nuclear, 184
dissonance among colleagues, 34
Dodds, Harold, 121, 122
    action against Bohm by, 135
DT burning, 116
DT mixture, 147
DT reaction, 110, 111, 157, 158
DuBridge, Lee, 20, 37n, 38
Du Pont, reactor construction by,
    104

Eames, Anthony, 189
Eddington, Arthur, 56, 56, 57
Egypt, 184n
Einstein, Albert, 47, 69
Eisenhower, Dwight, 181n
electric force, 60, 61, 62, 65
electromagnetism, laws of, 10
Elliott, Josephine, 106n, 148
Elugelab, 177, 178, 181n
$E = mc^2$, 47, 60
Emergency Capability Weapons,
    160
Energy, Department of, xiii
energy release
    in fusion reactions, 110, 111
Enewetak atoll, 112, 117, 177, 178, 182
    and Greenhouse tests, 108, 114
    and Ivy tests, 179
ENIAC, 103, 144, 150
equilibrium, thermal, 7
Evans, Cerda, 103
Evans, Dottie, 92
Evans, Foster, 103
Everett, Cornelius, 17, 87, 103, 104,
    105, 106
exclusion principle, 60
experiments, diagnostic, 115, 180
Exploratorium, 132
exponential change, 46

fallout, 114n, 160, 184
Family Committee, 20, 114–115, 147,
    155n

"Father of the H bomb," 13
F Division, 98
Fehner, Terry, 189
Fellows, Margaret, 141
Fermi, Enrico, 37n, 39, 40
    and 1941 idea of fusion, 96
    1945 lectures on the Super by,
        91, 98
    1949 meeting with Teller, 43
    and calculations with tritium,
        111
    and chain reaction, 93
    character of, 89
    early death of, 90
    GAC special role on, 40
    as head of F Division, 98
    at June 1951 meeting, 152
    as lecturer, 85
    as Los Alamos consultant, 87,
        88, 89, 119
    Super calculations by, 84, 103,
        105, 106
Feynman, Richard, 170–171
"first idea," 10–11
Fisk University, 132n
fission
    compared with fusion, 66
    contribution to H-bomb yield,
        158
    energetics of, 64
    energy released in, 66
    inhibition of, 65–66
    January 1939 news of, 53
fission bomb
    as source of X rays, 1
    as thermonuclear trigger, 156,
        164, 174, 180
fission physics, 59, 65–66
Fitzpatrick, Anne, 5n
Flagstaff, Arizona, 125
folk dancing, 25, 161
Ford, Adam, 190
Ford, Jason, 190
Ford, Joanne, 190
Ford, Kenneth, 141, 174

Ford, Masha, 189
Ford, Paul, 190
Ford, Sarah, 190
Ford, Star, 190
Forrestal, James, 130
Forrestal Campus, 130n
Forrestal Research Center, 130
    boiler room of, 169
France in nuclear club, 184
Frankel, Stanley, 93n, 94
Freeman, Burton (Burt), 85, 148
    shared optimism of, 148
    and "telephone book" report,
        107
Friden calculators, 5n
Frieman, Edward, 141
Frisch, Otto, 53–54, 54, 55
Froman, Darol, 17, 20
Fuchs, Klaus
    espionage by, 10, 12, 91, 100, 122
    and invention with von
        Neumann, 90–91, 99–101,
        111, 115, 116, 117
Furry, Wendell, 128
fusion
    at 1942 conference, 94
    compared with fission, 66
    energetics of, 64
    energy released in, 66
    inhibition of, 65
    in stars, 57
    of two deuterons, 66

GAC
    1949 composition of, 37n
    1949 main report, 37–38,
        39–40
    1949 majority annex, 38
    1949 minority annex, 39
    October 1949 meeting of,
        37–40
    September 1950 meeting of,
        106–108, 146
Gallup, New Mexico, 84
Gamma Building, 86

gamma particles, 50
gamma rays, 46
Garwin, Richard (Dick), 168n, 189
  designing Jugheads, 159–160
  designing Mike, 154–155
  as Los Alamos consultant, 119
  in T-DO, 87
  working with Rosenbluth, 159
Garwin design, 154–155, 156–158
Gedankenexperiment, 69–70
Geiger, Hans, 48, 49
General Advisory Committee. *See*
  GAC
General Atomics, 16, 168n
genocide and "super bomb," 38
George (test). *See* Greenhouse
  tests
Georgia, 71
German atomic program, 58n
Gignac, Savannah, 190
Ginzburg, Vitaly, 10, 11
Glazer, Tom, 92
Glennan, T. Keith, 152n
Goess, Robert, 141
Goudsmit, Samuel (Sam), 137
Grand Canyon, 125
Graves, Alvin, 179
Graves, Elizabeth, 182
graveyard shift, 151
Gravitation, 123
gravity, role of in Sun, 92
Great Britain in nuclear club, 184
Greenhouse tests, 14, 108, 114, 115
  Dog, 118
  Easy, 118
  experiments at, 115
  fission trigger used in, 174
  George, 114, 115, 117, 118, 143,
    147, 174
  George, early design of, 174
  George, success of, 116
  Item, 114, 131, 143
  Item, and high-energy
    neutrons, 147

Item, as test of boosting,
  117–118
Item, early design of, 148
Griggs, David (Dave), 182
Groves, Leslie, 97
  barring of Bohm by, 133
Guggenheim Foundation, 27
gun-type weapon, 99, 100

Hahn, Otto, 53
half life, 46
Hamburger Heaven, 151
Hamilton, Donald (Don), 121, 122
Hanford, Washington, 104
Hansen, Chuck, 189
Harrisburg, Pennsylvania, 186
H bomb, deliverable, 11
Heisenberg, Werner, 58n
helium-4, 63
Henyey, Louis, 128, 170
"Heterocatalytic Detonation," 1, 3
Hewitt, Paul, 189
hexadecimal notation, 163
Hill, David, 28, 29
Hitler, Adolf, 93
Hofstadter, Robert, 139n
Holt, Rush, 189
Hopkins, Matthew, 189
House Un-American Activities
  Committee. *See* HUAC
HUAC, 30, 122, 132, 134, 136
hydrogen "overweight," 56

IBM, 149
  in New York, 144, 151
implosion, radiation. *See* radiation
  implosion
independent-particle model, 28
India in nuclear club, 184
Institute for Advanced Study, 34,
  87, 120, 152, 169
Iran, 184
Ireland, 184n
iron, 64, 66
Iron Curtain, 122

Irvine, 186
Israel in nuclear club, 184
Item (test). *See* Greenhouse tests
"It's a boy," 182
Ivy Mike, 114, 178
See Also Mike

J Division, 179
Jemez Mountains, 88
"Joe 1," 10, 29, 30, 97, 102, 122
   Truman's announcement of, 37
"Joe 2" and "Joe 3," 79
John Carroll University, 72
Johnson, Louis, 42
Journal Club, 137–138
Jugheads, 159–160

kelvin (unit), 67
Kentucky, 71
Keyworth, George A. (Jay), 19
Kilburn, Tom, 165n
kiloton, defined, 1n
Konopinski, Emil, 93n, 95
   and 1941 idea, 96
Korean War, 80, 122

LAMS 1225, 1, 6, 18, 20, 154
Lamy, New Mexico, 144
Las Vegas, Nevada, 113
Lawrence, Ernest, 182
"layer cake," 5, 10–11, 12, 183
Layzer, David, 141
Legacy of Hiroshima, 15
Lennie, Frances, 189
Levin, Martin, 189
Libya, 184
Lilienthal, David, 41, 42
line of stability, 61–62
liquid-droplet model, 28
lithium deuteride, 22, 92, 102, 142
   calculational challenge of, 103, 111
   calculations with, 159, 168
   as "second idea," 11
   used in Bravo test, 161, 184

lithium hydride. *See* lithium deuteride
Livermore Laboratory, 102n
Lomanitz, G. Rossi (Ross), 132n, 135
Longmire, Conrad, 87
Los Alamos, New Mexico
   1968 summer in, 187
   airstrip, 76
   road to, 76
   schools, 120
   town entrance, 77
Los Alamos National Laboratory, 1n
Los Alamos Primer, 95
Los Alamos Ranch School, 77, 85
Los Alamos Scientific
   Laboratory, 1
   approves Matterhorn, 169
   buildings at, 85, 86
   and design of fission trigger, 143
   and formation of Matterhorn, 121
   and loyalty oath, 81–83
loyalty oaths, 81–83, 83n, 122
"Lucky Dragon No. 5," 114n, 160
"luminaries," 93

machine language, 167
Madison, Wisconsin, 131
magnetic monopole, 26, 124, 155–156
Manchester, UK, 48
Manhattan Engineer District, 97n
Manhattan Project, 97
MANIAC, 120, 144, 150, 165, 169
   memory of, 166
Manley, John, 82–83
   and 1942 conference, 93n
Marchant calculators, 5n
Mark, Carson, 2–3, 19, 87, 88
   1995 interview with, 21–23
   Bethe's link to, 88
   at June 1951 meeting, 152

and Nevada tests, 113
reaction to Wheeler-de
    Hoffmann meeting, 33
and Teller, 4, 20
and Ulam, 4, 17
Marsden, Ernest, 48, 49
Marshall, James, 97n
Marshall Islands, 112, 160, 177
mass-energy equivalence, 47
mass spectrometer, 55
Matterhorn, 102
    1952 team, 141
    approved by April 1951, 121
    beginning of, 140–142
    chosen location of, 130
    final report of, 168–169, 170,
        171–173, 172
    and Mike design, 159
    shack, 140
    work of, 163–176
Matterhorn (actual peak), 129
Mayer, Harris, 103, 189
McCarthy, Joseph, 30, 122
McGill University, 46
McIntosh, John, 141
McKibbin, Dorothy, 75, 76
McMahon, Brien, 41, 42, 43
McMillan, Edwin (Ed), 93n
mechanics, classical, 26
mechanics, laws of, 10
megaton, defined, 1n
Meitner, Lise, 53–54, 54, 55
mercury delay lines, 166
Metropolis, Nicholas (Nick), 103
Mexico, 184n
Mike, 111, 168
    as first big test of
        thermonuclear burning, 174
    as first Ivy test, 178
    fission trigger of, 175
    imagined design of, 158
    in its shed, 180
    mushroom cloud of, 181
    predicted yield of, 177
    sequence of events in, 177

time of explosion of, 178n
weight of, 179
yield of, 177, 183
Mississippi, 114
Monroe calculators, 5n
motion, molecular, 69
motorcycle, 118, 131, 161
Mottelson, Ben, 28n, 29n
Mt. Wilson telescope, 126
Murray, Peggy, 141
Murray, Thomas, 152n
mushroom cloud, 180, 181

Nason, David, 73
National Bureau of Standards, 144,
    163
Navy, 71–72
NBS, 144, 163
Nelson, Eldred, 93n, 94
neutrinos, 104
neutrons
    14-MeV, 110, 116, 118, 142, 158
    diffusion of, 94
Nevada test site, 113, 114
New Agenda Coalition, 184
New Mexico landscape, 74, 76
New Zealand, 184n
Nordheim, Lothar, 102, 119, 152
Noyes, Pierre, 172, 173
nuclear energy, 59–66
nuclear force, 60
nuclei, electric repulsion of,
    108–109

Oak Ridge National Laboratory,
    130, 138
oaths. See loyalty oaths
Obama, Barack, 184
Ojala, Audrey, 141
Old Pecos Trail, 75
Olesko, Kathy, 189
Oppenheimer, Frank, 132, 135
Oppenheimer, J. Robert, 37n, 38,
    39

and 1942 conference, 93, 96, 127
1943 meeting with Teller, 43
at 1954 hearing, 153–154
his "frustrated gratitude," 147
as GAC chairman, 106
GAC special role on, 40
at June 1951 meeting, 152–153
relation to Teller, 107
and September 1950 GAC meeting, 106–108
as wartime head of Project Y, 97
Otowi Mesa, 74

P. O. Box 451, 142
Packard, 72
Pakistan in nuclear club, 184
Palmer Physical Lab, 129, 135
paper tape input, 168
Parcheesi, 89
Parry Island, 179
Pasadena, California, 124
patents, 52, 53n, 90–91, 99
Pauli exclusion principle, 60
Pearl Harbor, 96–97
Pennington, Ralph, 141, 173
Pension Domecq, 30
photons, 68
Pike, Sumner, 41, 152n
"pile," 97
Planck, Miriam, 84–85, 105, 106, 148
plasma, 9, 100, 101
plasma physics, 134
plug boards, 149
plutonium, 59
plutonium-239, 110, 157
PM-B-37. See Matterhorn, final report of
polonium, 46
Pond, Ashley, 85
Princeton Junction, 143, 152
"Princeton physics," 89, 123
Princeton University, 55, 120

1950 Wheeler visit to, 34
approves Matterhorn, 169
graduate work at, 24–27
and military research, 121
programming, 148
Project Matterhorn. See Matterhorn
Project Y, 94, 97
Pupin Hall, 95

Q clearance, xiii, 79, 136
quadrupole moments, 28
Quakers, 186
qualifying exam, 24
quantum mechanics
course in, 134, 138
fundamentals of, 133, 135

Rabi, Isidor Isaac (I. I.), 37n, 39
radiation
energy of, 69, 70
from fission bomb, 8
identified by Becquerel, 44
pressure of, 70
as a substance, 8, 67–70
temperature of, 67, 68
in von Neumann-Fuchs invention, 91
See also cosmic background radiation
radiation implosion, 1–12 (Chapter 1), 96, 151
as Teller's idea, 9
in the USSR, 10
Radiation Laboratory, Berkeley, 102n, 182
"radio-actif" coinage, 45
radioactive fallout. See fallout
Radio-Activity, 46
radium, 46, 47
Rainwater, James, 29
Ranger, 113
reactors, 59
Reagan, Ronald, 19n
"Red scare," 135

Reiffel, Dorothea, 141
Reines, Frederick (Fred), 85, 104
Reines Raum, 85, 88, 89
Renault, 32
Rhodes, Richard, 158, 174n
Richards, Caroline, 190
Richtmyer, Robert, 5, 10
Rio Grande, 76
Rockefeller, Laurance, 130
Rockefeller Institute for Medical
    Research, 130
Rockefeller University, 130
Rongelap, 160
Rongerik, 160
Röntgen, Wilhelm, 45
Rose Bowl Parade, 124
Rosenbluth, Marshall, 159
    and design of Mike's fission
        trigger, 175
    use of SEAC by, 168
Route 66, 74
Route 285, 75
Rowe, Hartley, 37n, 38
runaway Super. See Super,
    runaway
Reupke, William (Bill), 189
Rutherford, Ernest, 46–49, 47
    and "moonshine" comment, 52

Sakharov, Andrei, 10, 11
    and Sloika ("layer cake"), 5, 10
Saletan, Gene, 25
Sandia Labs, 105, 144, 151
Sandstone, 112
Sangre de Cristo mountains, 74, 78
Santa Fe, New Mexico, 75, 76
    East Plaza Avenue, 75, 76
    Plaza, 75
Santa Fe Railroad, 144n
sausage (Mike shape), 156, 159,
    160, 180
    Garwin's first report on, 155
Savannah River, 104
Saxon, David, 81n
Schmitz, John, 186n

Schweber, Silvan (Sam), 25
Science Talent Search, 29n
Scientists Committee on Loyalty
    Problems. See SCLP
SCLP, 136–137, 185
Seaborg, Glenn, 15, 37n
SEAC, 163–164, 165, 166, 167, 170
"second idea," 10, 11
secrecy, xiii
Segrè, Emilio, 93n
Segrè, Gino, 189
seismic waves, 182
seismometer, 182
Serber, Robert, 91, 94
    and 1942 conference, 93n
Serber Says, 5
Shack, Christine, 141
Sheldon, John, 151
Shenstone, Allen, 35, 121
    his reaction to author's choice,
        35
    his reaction to Wheeler's
        choice, 35
Shepley, James, 14n
Shrimp, 160
Singer roadster, 165
six-day work week, 83–84, 101–102
Slaughter, Aimee, 190
"slingshot effect," 3
Sloika. See "layer cake"
Slovenia, 184n
Smith, Cyril Stanley, 37n, 38
Smyth, Henry (Harry), 41
    his call to Wheeler, 30
    at June 1951 meeting, 152n
    his support for Matterhorn, 121
Smyth Report, 30n
Socorro, New Mexico, 161
South Africa, 184, 184n
Soviet Union. See USSR
sparkplug, 157, 158
Spektor, Mariya (Masha), 189
Spitzer, Lyman, 126, 127
    and naming Matterhorn, 129,
        130

and SCLP, 136
spreadsheet on paper, 105
square dancing, 84, 161
St. Francis Cathedral, 76
St. George, Utah, 114n
St. Jean de Luz, 27, 32
Standing, Kenneth (Ken), 162
Stanford Linear Accelerator, 139n
Stanford University, 139n
Staver, Tor, 25
steady-state burning, 163, 164, 173
Stellarator, 130
Steward, Sarah, 189
stockpiles, 184
    United States, 112
Strassmann, Fritz, 53
Strauss, Lewis, 41, 42
strong force, 60
Sublette, Carey, 174, 183, 189
Sun
    central temperature of, 108
    energy generation in, 92
    surface temperature of, 67
Super
    coinage of, 96
    von Neumann-Fuchs version
        of, 90, 91, 99
Super, classical, 16, 68, 92–105
    (Chapter 9), 96, 143
    1946 conference on, 91
    as of summer 1950, 107–108
    Ulam's calculations on, 14
    Wheeler group calculations
        on, 146
Super, equilibrium, 1–12 (Chapter
    1), 8, 118, 153
    later accounts of, 13–23
        (Chapter 2)
Super, runaway, 6, 7, 12, 68, 98
Sweden, 184n
"Swiss cheese," 5, 148
Swordtail, 173
Szilard, Leo, 51
    and idea of chain reaction,
        50–52

and patent on chain reaction,
    52, 53n
and uranium chain reaction,
    93

Tannenwald, Nina, 190
Task Force 132, 178, 179
Taylor, Theodore (Ted), 87
T Division, 4, 14, 87
T-Division Office (T-DO), 87
tear gas, 186
"technically sweet," 152, 153
"telephone book," 107, 146, 147
teletype output, 163, 168
Teller, Edward, 3
    and 1941 idea, 96
    and 1942 conference, 93n
    1945 departure from Los
        Alamos, 101
    and 1949 GAC report, 43
    1949 trip east by, 43
    and 1950 report to GAC,
        106–108
    his account of meeting with
        Ulam, 21
    and April 1951 report, 156
    Bethe's link to, 88
    and boosting, 118
    character of, 3
    competitive spirit of, 89
    and compression, 7
    effect of "Joe 1" on, 30
    his faith in the young, 155–156
    as Family Committee chair, 114
    at June 1951 meeting, 152
    and LAMS 1225, 1
    and "layer cake," 10
    learning of Spitzer idea, 127
    and lithium deuteride, 11
    and Los Angeles APS meeting,
        124
    and Mark, 22
    meeting in Chicago, 144
    "observation" of Mike test, 182
    optimism of, 105, 108, 148

his participation in 1946 conference, 91
his plan to "observe" Mike test, 176
as recruiter of Rosenbluth, 168n
recruitment efforts by, 119
recruitment of author, 34–35
recruitment of Wheeler, 30, 31
relation to Oppenheimer, 107
relation to Ulam, 104–105
and selection of continental test site, 112
and selection of Savannah River site, 113
and six-day work week, 83
his support of Matterhorn, 121
in T Division, 87
and "telephone book" report, 107
and Ulam, 2, 4, 8, 13
his view of Truman statement, 43
and von Neumann-Fuchs invention, 91
and wartime work on fusion, 97–98
and Wheeler, 4
his wish to head T Division, 97
Teller, Mici, 4, 96, 120, 125, 161
Teller, Wendy, 5n
Teller-Ulam idea, 1–12 (Chapter 1), 67, 68, 142, 153, 182
compared with von Neumann-Fuchs invention, 101
later accounts of, 13–23 (Chapter 2)
reactions to at the time, 4
and related ideas at the time, 5
temperature of radiation, 68
testing weapons, 112–114
Theoretical Division. *See* T Division
Theoretical Megaton Group, 155
thermal energy, 65

thermonuclear burning calculations, 145
"third idea," 10, 11–12
Tinker Bell, 70
Toll, John, 33, 78, 79, 85, 102, 120, 141, 148, 161
administrative skills of, 141
as co-author of PM-B-37, 173
his decision to join project, 34
first days at Matterhorn, 140, 141, 169
in France, 27, 32
and interests in world affairs, 80
and Los Angeles APS meeting, 20, 123
and "Princeton physics," 86
programming by, 149, 165
returning from Los Alamos, 131
as roommate, 77, 79
shared optimism of, 148
and "telephone book" report, 107
and thermonuclear burning, 143
welcoming Spitzer, 127
ton (of energy), 1n
transmutation, 46
Trinity test, 112
trip
from Los Alamos, 131
to Los Alamos, 73–77
tritium, 6, 98, 104
Truman, Harry
briefing on Mike, 181n
December 1950 declaration of emergency by, 122
and Du Pont, 104
January 31, 1950 statement by, 31, 32, 41n, 42, 83, 101, 102, 122
receiving GAC report, 40
Tucson, Arizona, 161
TX-5, 174, 175

U. S. Navy, 71–72
"U. S. Nuclear Physics Reserve,"
173
UK. See Great Britain in nuclear
club
Ulam, Claire, 77
Ulam, Francoise, 18, 120
her account of discovery, 18–19
Ulam, Stanislaw (Stan), 2
his account of discovery, 15–17
and calculations with tritium,
110–111
and Carson Mark, 113
character of, 2
and early Super calculations,
103, 104, 105, 106
and equilibrium, 7
and Fermi's early Super
calculations, 84–85
friendship with von Neumann,
90
and Glenn Seaborg, 15–16
and LAMS 1225, 1
meeting with Teller, 21
participation in 1946
conference, 91
and Super calculations, 84–85
in T Division, 87
and Teller, 2, 4, 13
Ulams, Stan and Francoise, 77,
187n
unified model, 28n
UNIVAC, 170, 173
University of California, 81–82
as an inertial block, 121
University of Chicago, 87, 102
Met Lab, 123
University of Massachusetts, 83n
University of Michigan, 72
University of Minnesota, 132
University of Washington, 82, 83
uranium-235, 110
"Uranium Club," 58n

USSR, 12
50-megaton blast, 183
August 1953 thermonuclear
explosion, 183
first atomic explosion, 29
thermonuclear work in, 10
and weapons tests, 11
Utt, James, 186

V-12 program, 71–72
Van Vleck, John, 93n
Vienna, 156
Vietnam War, 185, 187
Villard, Paul, 46
von Neumann, John (Johnny), 19,
89
early death of, 90
friendship with Ulam, 90
at June 1951 meeting, 152
as Los Alamos consultant, 87,
88, 119
and MANIAC, 165
and Teller, 20
and work on ENIAC, 103
See also von Neumann-Fuchs
invention
von Neumann, Klari, 103
von Neumann-Fuchs invention,
90–91, 99–101, 111, 115, 116, 117

Walton, Ernest, 50
Washington, D.C., 186
weapons testing, 112–114
Weizsäcker, Friedrich von, 54
Weizsäcker mass formula, 54
Wellerstein, Alex, 189
Wellesley College, 72
Wells, H. G., 50, 59
Wheeler, Alison, 121
Wheeler, Janette, 85
in France, 31, 32
and Mici Teller, 4
moving to Los Alamos, 77
reaction to Los Alamos, 78, 120
trip to Los Angeles, 125, 161

Wheeler, John, 26, 72, 141
1949-50 in France, 27–28
1950 vacation with Janette, 32
1950 visit to Princeton, 34
autobiography of, 83
Bethe's link to, 88
Bohm, disappointment in, 134
Bohm, recruitment of, 133
and Bohr, 28–29
concern about underestimated
yield, 183
creating Matterhorn, 126
in crowded office, 85
decision to join project, 32
early days at Matterhorn, 140,
141, 169
and early Super calculations,
103
effect of "Joe 1" on, 30
joining Los Alamos, 102, 119
at June 1951 meeting, 152–153
and Los Angeles APS meeting,
20, 123, 124
as Matterhorn leader, 165
and Matterhorn recruitment,
128
as a mentor, 35
and naming Stellarator,
129–130
and nuclear deformation,
27–28
optimism of, 105, 108, 148
and "Princeton physics," 87
as principal author of PM-B-
37, 168, 173
reaction to Mike test, 181
recruitment efforts by, 119
seeking Princeton approval, 121

and six-day work week, 83–84
and Spitzer idea, 127
and Task Force 132.1, 179
as a teacher, 25–26, 27
and "telephone book" report,
107
and Teller, 4
and thermonuclear burning,
143
and Ulam, 4
witnessing Mike test, 176, 178
and "Yule log," 5
and the Zia Company, 86
Wheeler children
in France, 27, 29, 31, 32
in Los Alamos, 77, 78
and Los Alamos schools,
120–121
Whitworth, Fletcher, 189
Wilets, Lawrence (Larry), 25, 141,
173
Williams, Frederic C., 165n
Williams tubes, 165, 166
Winnipeg, Ontario, 162
Woodrow, Roy, 128, 129
"Work of Many People," 14
World War II, 93

yield of Mike
actual, 183
calculated, 173
and Swordtail calculations, 173
Yucca Flat, 113
"Yule log," 5

Zeldovich Yakov, 10
Zia Company, 86, 123

Printed in the United States
By Bookmasters

Printed in the United States
By Bookmasters